John Robert Sitlington Sterrett

An Epigraphical Journey in Asia Minor

Vol. II (1883 - 1884)

John Robert Sitlington Sterrett

An Epigraphical Journey in Asia Minor
Vol. II (1883 - 1884)

ISBN/EAN: 9783744746106

Printed in Europe, USA, Canada, Australia, Japan

Cover: Foto ©berggeist007 / pixelio.de

More available books at **www.hansebooks.com**

Archæological Institute of America.

PAPERS

OF THE

American School of Classical Studies at Athens

VOLUME II.

1883–1884.

An Epigraphical Journey in Asia Minor.

By J. R. Sitlington Sterrett, Ph.D.

BOSTON:
DAMRELL AND UPHAM.
1888.

NOTE.

————

THE second and third volumes of the Papers of the American
School of Classical Studies at Athens have been devoted to the
publication of the results of Dr. Sterrett's two journeys in Asia
Minor, made in the summers of 1884 and 1885. The third volume,
which was published in March, 1888, contains the report of the
Wolfe Expedition, made in 1885. The present volume is devoted
to the journey of 1884.

The Committee of Publication wish it to be distinctly understood,
that for obvious reasons, which they trust will commend themselves
to all, they have undertaken no editorial supervision of these volumes,
and that Dr. Sterrett is solely responsible for all that appears in
them under his name, as regards both the substance and the
form.

WILLIAM W. GOODWIN, } *Committee of*
THOMAS W. LUDLOW, } *Publication.*

June, 1888.

PREFACE.

—•◦•—

THE expenses of the journey in Asia Minor, the results of which are contained in this volume, were borne by myself, with the exception of one hundred and fifty dollars, which were contributed by gentlemen in Boston.

The inscriptions in whose headings no reference is made to a previous publication are new. Those in whose headings reference is made to some publication have been published before, but with inaccuracies.

The square brackets [ ] mean that what is inclosed between them was originally on the stone, but having become defaced has been supplied by me. The round brackets ( ) mean that what is inclosed between them was never on the stone, *i.e.* either that the word was abbreviated on the stone and has been written out in full, or else that an error of the stonecutter has been corrected by me. The broken brackets 〈 〉 mean that what is inclosed between them is on the stone, but that it is redundant.

The following Turkish terms need explanation : —

Ak, white.
Ashagha, lower.
Aghatch, a Tree.
Bash, a Head.
Bel, a Pass, generally low and broad; see *Gedik*.

Bunar, a living Spring; see *Puñar*.
Boghaz, literally a Throat, applied to defiles that lead up to a Pass (*Bel* or *Gedik*).
Böyük, large, big.
Dagh, a Mountain.

VI PREFACE.

Dere, a Valley, broad or narrow; applied also to Cañons.
Djami, a Mosque.
Düden, a Place where water sinks under the Ground; Καραβόθρα.
Eski, old.
Gedik, literally a Notch, applied to a Pass where the mountains rise up on both sides like a saddle; see *Bel*.
Gök, blue.
Göl, a Lake.
Hissar, a Castle.
Indje, narrow.
Irmak, a large River.
Kale, a Castle.
Kara, black.
Kassaba, a Market Town.
Kaya, a Rock.
Khan, a Caravansary.
Kieui, a Village.
Kilisse, a Church.
Kishla, Winter Quarters.
Kizil, red.
Köprü, a Bridge.
Kütchük, small.
Kuyu, a Well.
Medressi, a College for the Study of Law and Divinity.

Mesdjid, a small parish Mosque.
Monastir, a Christian Convent.
Ören, Ruins.
Orta, middle.
Ova, a Plain.
Puñar, a Variation of *Bunar*.
Sari, yellow.
Shehir, a Town.
Sivri, pointed, peaked; applied to sharp, abrupt mountain Peaks.
Su, literally Water; applied also to large Rivers.
Tash, a Stone.
Tchai, a small River.
Tcheshme, an artificial Fountain; see *Bunar*.
Tekke, a Mohammedan Convent.
Tepe, a Hill.
Toprak, Field, Soil.
Turbe, a Mausoleum or Chapel built over a Tomb.
Ulu, large.
Uzun, long.
Veran or *Viran*, Ruins, ancient Site.
Yaila, Summer Quarters.
Yaziülü, inscribed.
Yeni or *Yeñi*, new.
Yer, Earth, Dirt.
Yokara, upper.

I desire to tender again to Professor Heinrich Kiepert, of the University of Berlin, my most hearty thanks for the cartographical construction of my routes from observations and measurements made by me in the field.

The first part of my road-notes were turned over to Professor W. M. Ramsay, according to our agreement, by which the geographical results of that part of the journey during which we worked together were to belong to him, and the epigraphical results to me. Accordingly, my routes begin at Isparta, the point where I ceased to give my road-notes to Mr. Ramsay.

The routes made on the journey from Isparta to Ak Serai are laid down on the large map which accompanies the *Wolfe Expedition to Asia Minor.* The routes made on the journey from Ak Serai to the Euphrates, and from the Euphrates to Angora, are given in the two maps which accompany the present volume.

In editing this volume I have had suggestions from W. M. Ramsay, F. D. Allen, Th. Mommsen, B. Pick, and my lamented friend, J. McKeen Lewis.

J. R. SITLINGTON STERRETT.

June, 1888.

AN

EPIGRAPHICAL JOURNEY

IN ASIA MINOR,

DURING THE SUMMER OF 1884.

BY

J. R. SITLINGTON STERRETT.

AN

EPIGRAPHICAL JOURNEY IN ASIA MINOR.

————oo꒛ꭓoo————

IN the fall of 1883 I was in Smyrna, having just returned from my
summer's work with W. M. Ramsay, Esq., in Phrygia. I was making
preparations to return to the interior on a journey of my own, when
I received a telegram from Professor L. R. Packard, then Director
of the American School of Classical Studies at Athens, requesting me
to come to Athens immediately in order to assist him in the School.
I went to Athens at his call, but with the determination to indemnify
myself for the journey I had to abandon by undertaking a more
extended tour at my own expense during the summer of 1884.
Fortunately I was able to carry out my plans, and this present volume
embodies the results of that journey. Mr. Ramsay had also made
arrangements for spending this summer of 1884 in archæological
research in Asia Minor, and it seemed expedient for us to work in
concert as long as the general plan of our journeys would allow, for
thus a greater extent of country could be explored systematically.

In pursuance then of our agreement we met in Smyrna on May
15th, 1884, where I provided myself with the necessary travelling
outfit. I then went to Aïdin Giuzel Hissar, the ancient Tralleis, to
buy horses and make other final arrangements.

Mr. Ramsay, who was to be accompanied by A. H. Smith, Esq.,
of Cambridge, England, was detained in Smyrna, and in the mean-
time I undertook an excursion in the direction of Nazli, during which
I copied the first four inscriptions.

No. 1.

Kiösk. On a round pedestal in a café. It is broken at the top and bottom, there being some faint traces of a line at the bottom but none at the top. The Alpha bars vary as indicated.[1]

ΣΑΝΔΡΟΝΟΞΕΙΔΑΝ
ΝΕΙΚΟΜΗΔΕΑΒΙΟΛΟΓοΝ
ΑΞΙΟΝΕΙΚΗΝΔΙΑΤΕΤΗΝ
ΤΟΥΕΡΓΟΥΥΠΕΡΟΧΗΝΚΑΙ
5 ΤΟΚΟΞΜΙΟΝΤΟΥΗΘΟΥΣΝΕΙ
ΚΗΣΑΝΤΑΔΕΕΝΑΞΙΑΑΓΩΝΑΞ
ΙΗΕΝΛΥΚΙΑΔΕΚΑΙΠΑΜΦΥΛΙΑ
ΚΗ%ΒΟΥΛΕΥΤΗΝΔΕΑΝΤΙΟΧΕ
ΩΝΚΑΙΗΡΑΚΛΕΩΤΩΝΓΕΡΟΥ
10 ΞΙΑΞΤΗΝΔΕΜΕΙΛΙΣΙ^Ν

[Ἡ βουλὴ καὶ ὁ δῆμος
ἐτείμησεν Φλά(βιον) Ἀλέ-]
1 ξανδρον Ὀξείδαν
Νεικομηδέα, βιολόγον
Ἀσιονείκην διά τε τὴν
τοῦ ἔργου ὑπεροχὴν καὶ
5 τὸ κόσμιον τοῦ ἤθους, νει-
κήσαντα δὲ ἐν Ἀσίᾳ ἀγῶνας
ιη᾽, ἐν Λυκίᾳ δὲ καὶ Παμφυλίᾳ

---

[1] Ligatures occur: line 2, MH; line 3, HN bis; 4, HNK; 5, NE; 8, HN. In line 2 the O between Γ and N is small, as is also line 10 the Ω between I and N. In line 10 there was probably a horizontal bar connecting I and C; in other words, the two letters were HC in ligature, but this is conjecture, as I failed to see such a connecting bar, and my copy has IC as given above.

κϛ΄, βουλευτὴν δὲ Ἀντιοχέ-
ων καὶ Ἡρακλεωτῶν, γερου-
10 σιαστὴν δὲ Μειλ[η]σίων.

. . . .

This inscription is a replica of one found in 1866 in the theatre of Tralleis, and published by Waddington from a copy of Salvetti. The first two lines have been restored from the inscription of Tralleis [Le Bas-Waddington, *Voyage Archéologique*, 1652 *b*].

No. 2.

Kiösk. On a large round pedestal in the cemetery. A large segment has been broken out of the pedestal, and with it has disappeared the left side of the inscription. Cf. Le Bas-Waddington, Voyage Archéologique, 600 *a*. C.I.G. 2942 *d*.[1]

```
▨▨▨▨NEPΩNΛKΛΛ▨▨ΔION
▨▨▨▨ΣEBAΣTONΓEPMANIKON▨
▨▨▨▨AYTOKPATOPAΘEON

       ▨▨▨▨MOΞOKAIΣAPEΩNKAOIEPΩΣE
5               EΠIANΘYΠATOY
       ▨▨▨▨ΞPIOYΠΛA▨▨▨OYΣIΛOYAΞOYΔIΛNOY
                EΠIMEΛHΘENTOΣ
       ▨▨▨▨IBEPIOYKΛΛYΔIOYIEPOKΛEOYΣ
       ▨▨▨▨KYPEINAIEPOKΛEOYΣΦIΛOKAIΣAPOΣ
10     ▨▨▨▨ΞAΓNOY          YIOYΠOΛEOΣ
```

Νέρων[α] Κλ[αύ]διον
[Καίσαρα] Σεβαστὸν Γερμανικὸν
Αὐτοκράτορα θεὸν

---

[1] In line 7 end, ΔΙΛΝΟΥ is certain. In line 10, ΠΟΛΕΟΣ is certain, not ΠΟΛΕΩΣ.

[ὁ δῆ]μο[ς] ὁ Καισαρέων καθιέρωσε
5   ἐπὶ ἀνθυπάτου
[Τιβε]ρίου Πλα[ντίου Σιλονα[ν]οῦ (Αἰ)[λ]ι[α]νοῦ
ἐπιμεληθέντος
[Τ]ιβερίου Κλαυδίου, Ἱεροκλέους
[υἱοῦ], Κυρείνᾳ, Ἱεροκλέους Φιλοκαίσαρος
10   [Σ]άγνου?, υἱοῦ πόλε(ω)ς.

Mr. Waddington places this inscription in the last years of Nero's
reign, about 54 A.D.; cf. his commentary in *Voyage Archéologique*,
600 *a*.

No. 3.

*Kavakavak, near Kiösk.   Quadrangular stone built into the
wall of a well, with the inscription up.   Cf. Le Bas-Waddington,* Voyage Archéologique, 1652 *f*.

Π · Α Ι Λ Ι Ο Ν Λ Λ Κ Ι Π Α Λ
Τ Ο Ν Ε Π Ι Τ Ο Υ Κ Ο Π Ω Ν С
Α Υ Τ Ο Κ Ρ Α Τ Ο Ρ Ο Σ Α Δ Ρ Ι Α Ν
Κ Α Ι Σ Α Ρ Ο Σ Σ Ε Β Α Σ Τ
5   Π · Α Ι Λ Ι Ο Σ Π Λ Ο Υ Τ Α Γ Ο Ρ Ο Υ
Υ Ι Ο Σ Ε Ρ Μ Ο Δ Ω Ρ Ο Σ
Α Ρ Ε Τ Η Σ Ε Ν   Κ Α Ι Τ Η
Ε Ι Σ Τ Η Ν Γ   Ν ο Ι

Π. Αἴλιον ['Α]λκιπά[λην]
τὸν ἐπὶ τοῦ [κ]ο[ιτ]ῶν[ος]
Αὐτοκράτορος 'Αδριαν[οῦ]
Καίσαρος Σεβαστ[οῦ]
5   Π. Αἴλιος, Πλουταγόρου
υἱὸς, Ἑρμόδωρος
ἀρετῆς ἕ[νεκα] καὶ τῆ[ς]
εἰς τὴν [πόλιν εὐ]νοί[ας].

The reading of line 1 is certain. Mr. Waddington (loc. cit.) conjectures ΑΛΚΙΒΙΑΔΗΝ, and identifies him with the person mentioned *C.I.G.* 2947, 2948. The name Ἀλκιπάλης is certainly strange, but still not more so than many others that occur on Asiatic soil.

No. 4.

At a fountain by the roadside one hour west of Kiösk. It is a long rectangular stone, with a fragmentary inscription in two columns. The left end of the stone is broken away, and with it the commencement of the lines of the inscription forming Column I. The letters of this inscription are larger than those of its mate in Column II., which has been much worn away by the action of water. Cf. Le Bas-Waddington, Voyage Archéologique, 1652 ; *My* Preliminary Report, *p.* 4.

COLUMN I.

                ΟΣΤΟΥΕΝΤΗΙΕΡΑΚΩ
        ΚΕΝΑΞΙΩΜΑΔΙΟΥΕΛΕ
        ΙΞΙΟΝΙΕΡΑΞΚΟΜΗΞΚΑΤΟΙ
        ΙΔΡΥΜΕΝΑΤΩΑΠΟΛΛΩΝΙ
    ΣΤΑΞΤΟΥΘΕΟΥΘΕΡΑΠΕΙΑΕ
        ΞΑΠΑΡΧΗΞΕΙΧΕΝΕΓΩΔΕ
        ΠΟΤΩΝΠΡΟΕΜΟΥΒΑΞΙ
        ΞΕΙΝΤΕΚΑΙΤΑΤΩΝΘΕ
                                ΘΗΝ

COLUMN II.

ΤΕΚΑΙΩΞΕΤΙΜΗΘΗΔΙΑΤΑ
ΤΗΝΠΑΤΡΙΟΝΒΑΛΕΙΑΝΚΑ
ΤΟΞΤΕΤΑΥΠΟΤ
ΞΚΗΠΤΡΟΝΕΧΟΥ ΗΙΚ
ΥΝΤ ΛΕΙΝΚΑΘΥΔ

COLUMN I.

. . . .

. . . . ος τοῦ ἐν τῇ Ἱερᾷ Κώ[μῃ]

. . . . . . [ἔ]δωκεν ἀξίωμα δι' οὗ ἐλε-

[υθέρους ἀφῆκε τοὺς πλη]σίον Ἱερᾶς Κ[ώ]μης κατοι-

[κοῦντας καὶ τὰ] ἱδρύμενα τῷ Ἀπόλλωνι

. . . . . . [εἰ]ς τὰς τοῦ θεοῦ θεραπεί[ας]

. . . . . . [καθὼ]ς ἀπ' ἀρχῆς εἶχεν· ἐγὼ δε

. . . . . . [ἀ]πὸ τῶν πρὸ ἐμοῦ βασι-

[λέων . . . . . αὐξ?]εω τε καὶ τὰ τῶν θε-

[ῶν] . . . . τὴν.

COLUMN II.

τε καὶ ὡς ἐτιμήθη διὰ τα[ύτης τῆς ἐπιστολῆς?]

τὴν πάτριον βασιλείαν κ[αὶ] . . . .

τός τε τὰ ὑποτ[εταγμένα] . . . .

σκῆπτρον ἐχούσ[ῃ] . . . .

[σ]υντ[ε]λεῖν κάθυδ[ρον]? . . . .

. . . .

This inscription is a fragment of a letter of one of the later kings, possibly Antiochus the Great, in regard to the people of Hiera Kome and the sanctuary of Apollo.

At Kuyudjak I met Messrs. Ramsay and Smith. From this point our final start was made, going by way of Antiochia to Aphrodisias, the modern Geira.

Antiochia has disappeared entirely, it seems, and from the villages of this region we collected only a few insignificant inscriptions.

No. 5.

Ali Agha Tchiftlik. On a square marble basis. Circular anathema with a hole in the centre. Copied by W. M. Ramsay.[1]

```
▨▨▨▨▨▨▨▨Ϲ Κ Α Τ Η Ν
▨▨▨▨▨▨▨Ν Χ Α Ρ Μ Ι
▨▨▨▨▨▨Ͻ Ν  [blank space]
▨▨▨▨▨ Ν Κ Ρ Ά Τ Ι Ο Ν
5   ▨▨▨▨Ϛ Α Μ Ε Ν Ο Υ Τ Η Ϛ·
▨▨▨▨Α Ϛ Ε Ω Ϛ Τ Ο Υ Α Ν Δ Ρ Ι
▨▨▨▨Ζ Ω Ϛ Ι Μ Ο Υ Τ Ο Υ ⅘
▨▨▨▨Ρ Ο Ϛ Α Υ Τ Ο Υ  [blank]
▨▨▨▨Ο Υ Τ Ω Ν Π Α· Τ Ρ Ι Ω Ν
10  ▨▨▨▨Λ Ι Τ Ο Υ Κ Υ Ρ Ι Ο Υ
▨▨▨▨Τ Ο Κ Ρ Α Τ Ο Ρ Ο Ϛ
```

· · · · · · · · · ·

...... Χαρμί-
[δην παίδ]ων
[νικήσαντα? πα]νκράτιον·
5 [προνοη]σαμένου τῆς
[ἀναστ]άσεως τοῦ ἀνδρι-
[άντος] Ζωσίμου τοῦ
[πατ]ρὸς αὐτοῦ,
[νεωκόρ]ου τῶν πατρίων
10 [θεῶν κα]ὶ τοῦ κυρίου
[αὐ]τοκράτορος.

---

[1] The lower lines are 10¾ inches long; the lost space is 5¼ inches. In line 1, HN are in ligature.

No. 6.

*Ali Agha Tchiftlik. In a cemetery on a hill near a Turbe,
about fifteen minutes east of the village.*

T O M N H M E I O N	τὸ μνημεῖον
Λ Π Ο Λ Λ Ω Ν Ι Ο Υ Τ Ο Υ	['A]πολλωνίου τοῦ
Λ Π Ο Λ Λ Ω Ν Ι Ο Υ	['A]πολλωνίου.
Z H	Ζῇ.

No. 7.

*Ali Aghi Tchiftlik. Broken at both ends ; letters six
inches high. Copied by W. M. Ramsay.*[1]

Ⅳ Φ ▨ Ξ Ι Λ Ι Ν Γ▨

No. 8.

*Yeñidje. Large block broken at both ends, now serving as
a mouth-piece to a well.*

▨Λ Ρ Ι Ω Ν Α Γ Α Ι Ο Υ▨
▨▨Π Ε Ρ ⋮ Π Ο Π▨

. . . Θεα]ρίωνα? Γαΐου
['Aσ]περ? Πόπλ[ιος?].

A large number of inscriptions from Geïra (Aphrodisias) have
been published already, and consequently we could not hope for
great epigraphical gain unless we should spend a number of days
among the ruins, in order to sift the new from the old, the unknown
from the known inscriptions. But time pressed, and we reluctantly
abandoned the plan of investigating the site carefully. Still our visit
was not wholly without fruit.

---

[1] NΓ are in ligature.

No. 9.

Geira [*Aphrodisias*]. *Slab with mouldings: length inside the mouldings, 0.75 m. (including the moulding, 0.88 m.); height within the moulding, 0.55 m. (including moulding, 0.69 m).
Copied by J. R. S. S.; copy verified by W. M. Ramsay.*

```
▨ΙΑΤΟΥΠΑΤΠΙΟΥ▨ΙϹΗΝϹΟΡΟΝΚΕΚΗΔΕΥΤΑΙΘΕΟΔΟΤΟ
   ϹΑΥΤΟϹΚΑΙΑΜΜΙΑ·ΑΡΙϹΤΕΙΔΟΥΤΟΥΖΗΝΩΝΟϹ
   ΙΑΝΑΝΟΙΖΑΙΤΗΝϹΟΡΟΝΜΕΤΑΤΟΕΝΤΑΦΗΝΑΙ
   ΝΧΩΡΗϹΙΝΗΟΙΚΟΝΟΜΙΑΝΤΙΝΑΠΕΡΙΤΗϹ
   ΝΤΙΠΟΙΗϹΑϹΩΔΗΠΟΤΕΤΡΟΠΩϹ
   ΜΑϹΤΩΝϹΕΒΑΤΩΝΑΡΓΥΡΙΟΥ✳ϚΗ
   ΙΤΗΥΤΟΤΗΝϹΟΡΟΝΚΗΔΕΥΘΗϹΕΤΑΙΔΙΟ
   ΝΟϹΟΤΟΥΑΔΕΛΦΟΥΜΟΥΥΙΟϹΕΤΕΡΟϹ
   ΙΟΥΚΑΙΟΙΕΖΑΥΤΩΝΓΕΝΟΜΕΝΟΙΠΕΡΙ
   ΑΝΤΟΙΗϹΑΙΟΥΔΕΝΙΤΡΟΠΩϹΤΕΖΑΛ
   ΟΙϹΕΤΑΝΩΓΕΓΡΑΜΜΕΝΟΙϹϹΠΡΟϹΤΕΙ
   ΚΙΟΝΕΠΙϹΤΕΦΑΝΗΦΟΡΟΥΤΙΒΕΡΙΟΥ
```

[. . . . . .᾿Αμμία τοῦ Παπίου, εἰς ἣν σορὸν κεκήδευται Θεόδοτο-
[ς, κηδευθήσεται δὲ . . . .]ς αὐτὸς καὶ ᾿Αμμία ᾿Αριστείδου τοῦ Ζήνωνος·
[ἕτερος δὲ οὐδεὶς ἕξει ἐξουσ]ίαν ἀνοῖξαι τὴν σορὸν μετὰ τὸ ἐνταφῆναι
[τοὺς προγεγραμμένους πάντας, οὐδὲ συν]χχώρησιν ἢ οἰκονομίαν τινὰ περὶ τῆς
[σοροῦ . . . . ἐπεὶ ὁ ἀ]ντιποιήσας ᾧ δήποτε πρόπῳ ἔσ-
[ται ἀσεβὴς καὶ τυμβωρύχος καὶ ἀποτείσει εἰς τε]ι̣μὰς τῶν Σεβαστῶν ἀργυρίου * ͵ς
[. . . . . . . ἐν δὲ τῇ εἰσώστῃ]? τῇ ὑπὸ τὴν σορὸν κηδευθήσεται Διο-
[γένης? . . . .]νος ὁ τοῦ ἀδελφοῦ μου υἱὸς· ἕτερος
[δὲ οὐδεὶς . . . .]μον καὶ οἱ ἐξ αὐτῶν γενόμενοι περὶ
[. . . . .]αν ποιῆσαι οὐδενὶ τρόπῳ ἐπεξαλ-
[λοτριῶσαι? τὸν πλάταν . . . . . .τ]οῖς ἐπάνω γεγραμμένους προστεί-
[σει] τούτου ἀντίγραφον ἀπετέθη εἰς τὸ Χρεωφυλά]κιον ἐπὶ στεφανηφόρου Τιβερίου
[Κλαυδίου Τημικλέους].

No. 10.

Geïra.   Inscription on a large stone, circa 6 × 4½ *feet.
Letters ornamentally cut,* 1⅝ *inches in height.   Copied by
A. H. Smith.*

```
O R Ω M O Σ K A I H E Π I K E I M
Y Π O T I B E P I O Y I O Y Λ I O Y
I O Λ I A N O Y   Χ   E I Σ H N Σ O P
R I A N O Σ K A I O Y Λ Π I A ·   K
5 A Y T O Y E T E P O Σ Δ E O Y
A Y T H N ⌐ E A N Δ E T I Σ E Π
T Ω N E N K H Δ E Y O H▨▨▨Σ C
T Ω N Δ I A T E T A Γ M E▨▨N ⌐
```

[. . . . καὶ] ὁ [β]ωμὸς καὶ ἡ ἐπικειμ[ένη αὐτῳ σορὸς]
[κατεσκευάσθησαν?] ὑπὸ Τιβερίου Ἰουλίου[. . . . . .]
[. . . . . .] Ἰουλιανοῦ, εἰς ἣν σορ[ὸν κηδευθήσονται]
[αὐτὸς καὶ Οὐαλε]ριανὸς καὶ Οὐλπία κ[αὶ οὓς ἂν
   βουληθῶσιν]
[αὐτὸς? καὶ ἡ δεῖνα ἡ γυνὴ?] αὐτοῦ, ἕτερος δὲ οὐ[δεὶς
   ἕξει ἐξουσίαν]
[ἐνθάψαι τινὰ ἄλλον εἰς] αὐτήν· ἐὰν δέ τις ἐπ[εισβιάζηται]
[. . . . .]των ἐνκηδευ[θ]η[. . . . .]
[. . . . .]τῶν διατεταγμέ[νων . . . .]

From Geïra Messrs. Ramsay and Smith went around Baba Dagh
to the north, by way of Deñizli, and I to the south.   On this excursion these gentlemen copied the following two inscriptions.

No. 11.

Assar.   In a wall.   Copied by W. M. Ramsay.

```
▨▨▨▨▨Δ Ρ Α Ξ Ι▨▨
▨▨▨▨▨Ο Υ Τ Ο Υ Λ̅ξ̅
▨▨▨▨▨≋ Ω Τ Λ Τ Ο
▨▨▨▨▨Ͽ Ε Τ Ο Υ Δ
5 ▨▨▨▨▨Ι Ο Υ
▨▨▨▨▨Λ Ι Π Τ Η Ν
▨▨▨▨▨Ε Ι Ν Ι Α Ν Ο Ϝ
▨▨▨▨▨Χ Ρ Ο Ν Ι Ο Υ
```

. . . . . ἀν]δράσι? . .

. . . . . . τοῦ . .

. . . . δοκιμ]ωτ[ά]το[υ]?

. . . ἀγωνο]θέτου δ-

[ιὰ β]ίου

. . . . . [ἀ]λίπτην?

. . . Λονγ]εινιανὸ[ς]

. . . . χρονίου.

No. 12.

*Hadji Eyuplu, half an hour from Deñizli.   Copied by
W. M. Ramsay.*

The inscription is on a stele with a gable, in which is represented
the sun; below the gable is inscription *A*.   Below this is an arched
niche, in which are represented two human figures.   On the arch is
inscription *B*.

A.

Z ⏝ Ƨ A Δ I M O Λ O Ƨ ⏝ O X ⏝ P O Ƨ
O K I Λ A P A Z E ⏝ N M N I A Ƨ X A P I N

B.[1]

Ė Λ Π I Ƨ Π A P O Δ I T A I Ƨ X E P I N

A.

Ζωσᾶδι Μολοσῷ ὁ χῶρος
ὁ Κιλαραζέων μνίας χάριν.

"The country of the Kilarazeis to Zosas Molosos, by way of
remembrance."

B.

Ἐλπὶς παροδίταις χέριν.

" Elpis greets the passers-by."

The name Ζωσᾶς occurs *C.I.G.* 3665, but neither is this form or
the form Σωσᾶς, — ᾶτος common in Greek onomatology (see *Revue
Archéologique*, 1878, XXXVI. p. 318, and Letronne, *Inscriptions
Grecques et Romaines de l'Egypte*, II. p. 457.

Possibly the ZⓌƧAΔI of our inscription may be a mistake for
ZⓌƧATI or ZⓌƧAΔH. The form Σωσάδης occurs in an inscription
of Athens in Φιλίστωρ III. p. 568. ₀

May 29. Geïra to Makuf, 4 h. 40 m. The plateau upon which
Aphrodisias was situated contracts gradually as one advances, until it
strikes the foot of a spur of Baba Dagh immediately beyond Besh
Kavaklar. We cross this spur of Baba Dagh, and in 2 h. 15 m. from
Besh Kavaklar we reach its foot in the Davas Ova. Traversing the
plain we reach Kara Hissar in 35 m.

---

[1] In line 3, XEPIN stands for XAIPEIN.

No. 13.

Kara Hissar. Block now used as a mouth-piece to a well near the village. Length, 1.10 *m.; width,* 0.90 *m.; height of letters,* 0.06 *m.*

⬛⬛⬛⬛⬛⬛⬛ ⋚ T O N ⋚ E ⬛⬛⬛⬛⬛⬛⬛
⬛⬛⬛⬛⟨ O N A P X I E P E A M E Γ I ⋚ T C ⬛⬛⬛
⬛⬛⬛) Y ⋚ I A ⋚ T O 🔲 ⬛⬛ ⊼ C K P A T O P I ⬛⬛
⬛⬛⬛ T O Ï Π A ⸗ ⬛⬛⬛⬛⬛⬛⬛ I Δ O⸗⬛
⬛⬛⬛ Λ N O Y ⋚ Λ ⬛⬛⬛⬛⬛⬛⬛⬛⬛
⬛⬛⬛⟨ Δ I A Θ H K H ⬛⬛⬛⬛⬛⋚ T ⊼ ⟨ ⟩ ⬛
⬛⬛⬛ Y Γ E N O M ⟨⟩⟩ ⸱⸱ Ö ⟨ A P X I A T ⬛⬛
⬛⬛⬛ Ⳅ Λ ↑ Λ ⌒⸱⸱⸱ T O Y K Y P I O Y K ⬛⬛⬛

. . . .[ἄρισ]τον? Σε[βαστὸν] . . . .
. . . . . [Γερμανι?κ]ὸν ἀρχιερέα μέγιστ[ον] . . . .
[δημαρχικῆς ἐξο]υσίας τὸ [ιθ′, αὐτο]κράτορ[α]
[τὸ -? ὕπατον] τὸ ζ′, πα[τέρα πατρίδο[ς], . . .
. . . . . . ανουσα . . . .
. . . . [ἐκ] διαθήκη[ς Τίτου?] Στ[ατιλίου]
. . . . ου γενομ[ένου] ἀρχιάτ[ροῦ καὶ]
[στεφανηφόρου] τοῦ κυρίου Κ[αίσαρος].

Concerning the ἀρχίατρος, see Marquardt, *Privatleben*, II. p. 755, No. 4 ; Le Bas-Waddington, *Voyage Archéologique*, 1695 ; *C.I.G.* 3953 h ; *Bulletin de Correspondance Hellénique*, 1883, p. 360, 1885, p. 337, No. 20.

The office of στεφανηφόρος is connected with that of the ἀρχίατρος in an inscription of Heraclea given in *Bulletin de Correspondance Hellénique*, 1885, p. 337, No. 20, so that it must probably be restored here.

Travelling east from Kara Hissar we reach Makuf, the site of the ancient Heraclea (see Le Bas-Waddington, *Voyage Archéologique*, 1695, and *Bulletin de Correspondance Hellénique*, 1885, p. 330), in 22 m.

The Stadion at Heraclea is still very distinct. The Acropolis is a low hill of great extent on top. The walls of the Acropolis are

easily followed around the whole circuit. In some places they are
level with the ground, while in others they are still erect. The walls
have been destroyed and then rebuilt, as is clear from the architectural fragments, and even inscribed stones which are built into the
present wall. But that the foundations of the wall date from a comparatively early period is shown by the fact that on the outside the
wall is provided with finely executed stone shoots at the bottom to
carry the water off. Still, it must be noted that, at a place where the
wall is now used as a quarry by the villagers of Makuf, I discovered
an honorary inscription (No. 15) in the very foundation. The walls
were evidently rebuilt in time of great and pressing need, when the
anxious citizens made use of anything in the shape of stone that came
in their way.

No. 14.

Makuf [*Heraclea*]. *Near the Acropolis walls and close to the
Stadion. The stone is unpolished and very roughly hewn.
See my* Preliminary Report, *pp.* 4, 5. *Shortly after its
appearance in the* Preliminary Report *the inscription was
also published in the* Bulletin de Correspondance Hellénique,
1885, *p.* 332. *I had to copy the inscription in a rain and
could not read the last lines given by the French gentlemen,
who saw the stone under more propitious circumstances. It
is* 0.41 *m. in height;* 0.50 *m. in width.*

```
HΘHKHHΓOPACΘHYΠOTITOYCTATIAIὓ////
MHTIOXOYENHTEΘHCETEAYTOCKAIHΓYNɩ///
AYTOYAYPHΛIAMEΛITINHΔIONYCIOYK////
ONANAYTOIΠEPIONTECBOYΛHΘWCINETE
5 PWΔEOYΔENIEΞECTAIENΘAYETINAEITC//
  ἑ)ENΘAYAITINAΠOTICEITWKYPIAKW
  ℘ICKW✳ΦKAITHBOYΛHTHHPAKΛEW
  TWN✳ΦKEOYΔENḢTTONOENTAϹO
  ///IETAT///ΘH////////////////
10 //////////OYNANTIΓPAΦONAΠE///
  //////EICTAAPXEIA////////////
  ///////OCE////////////////////
```

Ἡ θήκη ἠγοράσθη ὑπὸ Τίτου Στατιλί[ου]
Μητιόχου, ἐν ᾗ τεθήσετε(=αι) αὐτὸς καὶ ἡ γυν[ὴ]
αὐτοῦ Αὐρηλία Μελιτινὴ Διονυσίου κ[αὶ]
ὃν ἂν αὐτοὶ περιόντες βουληθῶσιν· ἐτ[έ]-
5 ρῳ δὲ οὐδενὶ ἔξεσται ἐνθάψε(=αι) τινά· ἐ[πεὶ]
[ὁ] ἐνθάψα(ς) τι(ὰ) ἀποτίσει τῷ κυριακῷ
[φ]ίσκῳ (δηνάρια πεντακόσια) καὶ τῇ βουλῇ τῇ
Ἡρακλεω-
τῶν (δηνάρια πεντακόσια), κὲ οὐδὲν ἧττον ὁ ἐντα[φεὶς]
[μ]ετατε[ε]θή[σετε(=αι)· τῆς ἐπιγραφῆς ταύ]-
10 [της ἁπλ]οῦν ἀντίγραφον ἀπε[τέ-
[θη] εἰς τὰ ἀρχεῖα, [ἔτους] . . . .
[μην]ὸς ἕ[κτου, ἡμέρᾳ] . . . . . .

Line 3. The *Bulletin* reads ΑΥΡΗΛΙΑΙ instead of ΑΥΡΗΛΙΑ.
Line 5. The *Bulletin* reads ΕΝΤΑΨΕ for ΕΝΘΑΨΕ; and in
line 6, ΕΝΤΑΨΑΣ instead of ΕΝΘΑΨΑΣ. On the contrary, the
reading of the *Bulletin* at the end of line 5, ΕΠΕ is certainly more
accurate than my ΕΙΤΣ.

No. 15.

*Makuf.   Quadrangular cippus in the wall of the Acropolis.
Long, 1.30 m.; wide, 0.45 m.*

```
▨▨ΟΥΛΗΚΑΙ▨▨▨▨
▨ΤΕΙΜΗΣΑΝΙΕΡΩΝ▨▨ΑΜΕ
ΝΕΣΘΕΩΣΠΡΥΤΑΝΙΝΚ▨▨
ΣΤΕΦΑΝΗΦΟΡΟΝΚΑΙΓ▨
5 ΜΝΑΣΙΑΡΧΟΝΚΑΙΑΓ▨ ·
ΝΟΘΕΤΙΝΕΚΤΩΝΚΑ▨
ΛΕΙΦΘΕΝΤΩΝΤΗΠΟΛΕΙ
ΥΠΟΑΠΟΛΛΩΝΙΟΥΤΟΥ
ΤΥΔΕΩΣΤΟΥΑΝΔΡΟΣ▨
10 ΤΗΣΚΑΘΑΔΙΕΤΑΞΑΤ▨
ΟΑΠΟΛΛΩΝΙΟΣΤΗΝΕΠΙ
```

ΜΕΛΙΑΝΤΗϹΑΝΑϹΤΑϹΕ
ΩϹΠΟΙΗϹΑΜΕΝΩΝΞΚΥ
ΜΝΟΥΚΑΙΑΠΟΛΛΟ
15 ΦΑΝΟΥϹΤΩΝΛΔΡΑϹ
ΤΟΥϹΚΥΜΝΟΥΑΓΩ
ΝΟΘΕΤΩΝΤΗϹΑΓ▨▨▨
ΤΑΕΤΗΡΙΔΟϹ

['Η β]ουλὴ καὶ [ὁ δῆμος
ἐ]τείμησαν Ἱερω[νίδ]α Με-
νεσθέως πρύτανιν κ[αὶ]
στεφανηφόρον καὶ γ[υ]-
5 μνασίαρχον καὶ ἀγ[ω]-
νοθέτιν ἐκ τῶν κα[τα]-
λειφθέντων τῇ πόλει
ὑπὸ Ἀπολλωνίου τοῦ
Τυδέως τοῦ ἀνδρὸ[ς αὐ]-
10 τῆς, καθ᾽ ἃ διετάξατ[ο]
ὁ Ἀπολλώνιος· τὴν ἐπι-
μέλιαν τῆς ἀναστάσε-
ως ποιησαμένων Σκύ-
μνου καὶ Ἀπολλο-
15 φάνους τῶν ['Α]δράσ-
του Σκύμνου ἀγ[ω]-
νοθετῶν τῆς (ὀγδόης) [πεν]-
ταετηρίδος.

Two similar inscriptions from Makuf have been published in the *Bulletin de Correspondance Hellénique*, 1885, pp. 338–339, one of which is in honor of Hieronis, and the other in honor of Apollonios himself.

Concerning the conferring of honors, such as those mentioned in this inscription, upon women, see *C.I.G.* 3415, 3953 c and d; Curtius, *Beiträge zur Geschichte und Topographie Kleinasiens*, p. 62; *Bulletin de Correspondance Hellénique*, 1885, p. 339; *Journal of Philology*, XI. p. 143.

No. 16.

*Makuf.   Cippus lying by the side of the Acropolis walls.
Length, 1.27 m.; width, 0.33 m.*

```
                  H.MOΣETIMHΣAN
                  PYΦΩNOΣYIONHPΩA
                  ΣANTAΔIOΛOYTOY
                  ΣAΣAΛEYTΛΣH
 5                KAINYKTOΣΠPΩ
                  ΛTETHNIΔI
                  AΣTΩNΠIO
                  ΠATPIΔAEY
                  ΣXEΣINKAI
10                MNAΣIAPXIAΣ
                  NAΘEΣINTOY
                  IΣΛMENHΣTATI
                  'ΘΓATPOΣIEPEIAΣ
                  ΔIKAIOΣYNHΣTHΣ
15                ETOYΣ—HNP.
```

['Η βουλὴ καὶ ὁ δ]ῆμος ἐτίμησαν
['Ατταλον Τ]ρύφωνος υἱὸν ἥρωα
[ἀγορανομή]σαντα δι' ὅλου τοῦ
[ἔτους . . . .
5 . . . . καὶ νυκτὸς πρω-
. . . . τε τὴν ἰδί-
[αν . . . . [τῶν Πιο-
[νιτῶν . . . . . πατρίδα, εὐ-
[σχημόνως ζήσαντα . . .]σχέσιν καὶ
10 . . . . γυ]μνασιαρχίας
. . . . την ἀνάθεσιν τοῦ
[ἀνδριάντος ποιη]σ[α]μένης Τατί-
[ας, 'Ατταλου] θ[υ]γατρὸς, ἱερείας
. . . . δικαιοσύνης τῆς
15 πρὸς τὴν πόλιν]. ἔτους ηνρ'.

In line 15, the units come first, as is the case in Nos. 19 and 26.
If the era used be that of *Sulla*, then the inscription dates from
the year 74 A.D.; if the era be that of *Cibyra*, then the date is
183 A.D.

No. 17.

*Makuf. By the side of the walls. Greatest height, 0.50 m.;
width, 0.47 m. Cf.* Bulletin de Correspondance Hellénique,
1885, *p.* 337.

```
        M I Δ O Y M E N /%
        Y T A N I N K A I Σ T %
        Φ O P O N K A I A P X I A
        Ч A T Ω N E Y Γ E N E Σ
   5      , Ω N K A I E Y Σ X H M O N E Σ
        I A T Ω N A Π O Π P O Γ O N Ω N B O Y
        Λ E Y T Ω N Π A Σ A Σ A P X A Σ T %
        K A I Λ E I T O Y P Γ I A Σ E K T %
        K O T A T H Π A T P I Δ I K A %
  10    Λ A N Π P O T A T O N Γ A %
        M O T A T O N Θ Y Σ A I %
        Π A T P I O Σ Θ %
        Ь A Σ T Q %
```

        . . . . [τ]-
        [οὗ Χαρ]μίδου Μεν[άν]-
        [δρου, πρ]ύτανιν καὶ στ[ε]-
        [φανηφ]όρον καὶ ἀρχία-
        [τρον, ἔν]α τῶν εὐγενεσ-
    5   [τάτ]ων καὶ εὐσχημονεσ-
        [τ]άτων ἀπὸ προγόνων βου-
        λευτῶν, πάσας ἀρχάς τ[ε]
        καὶ λειτουργίας ἐκτ[ετελε-]?
        κότα τῇ πατρίδι κα[ὶ ἐπὶ τὸ]?
   10   λανπρότατον (κ)α[ὶ πολυδαπα-]?
        (ν)ότατον θύσα[ντα τοῖς]
        πατρίοις θ[εοῖς καὶ τοῖς Σε]-
        βαστ[οῖς] . . . .

No. 18.

Makuf. In the wall of the so-called Kalc. Two panels side by side on the same stone. The left panel is broken through the middle of the inscription. The right panel has been published in the Bulletin de Correspondance Hellénique, 1885, p. 341.[1]

A.

ΟΝΚΑΤΕΧΕΙ
ΟΝΕΝΤΤΟΡΟΝΟΥΤ
ΥΝΒΟΛΟΓΔΟΙΚΟΝ
ΗΤΕΚΝΩΜΙΓΑΚΑΙ
ΩΜΕΥΝΩΚΑΙΔΙΑΤΗϹ
ΛΛΗϹΧΑΙΡΕΛΕΓΕΙ
ΑΡΟΔΟΙΙ

B.

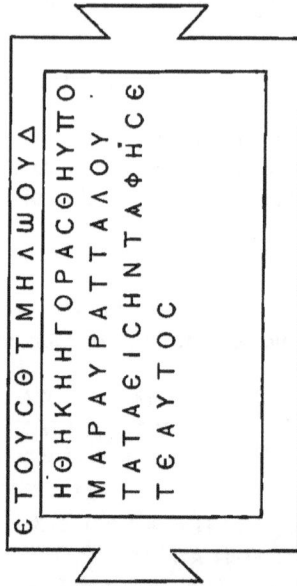

ЄΤΟΥϹΘΤΜΗΛШΟΥΔ
ΘΘΗΚΗΗΓΟΡΑϹΘΗΥΤΟ
ΜΑΡΑΥΡΑΤΤΑΛΟΥ·
ΤΑΤΑЄΙϹΗΝΤΑΦΗϹЄ
ΤЄΑΥΤΟϹ

[1] Ligatures occur in *B*: line 1, MH; line 2, HK, HHГ.

A.

_ ⌣ ⌣]ον κατέχει [κλυ]τὸν ἔνπορον οὗτ[ος ὁ τ]ύνβος
ὅ[ς] δ᾽ οἶκον ⌣ ⌣ η τέκνῳ μίγα καὶ [συνο]μεύνῳ,
καὶ διὰ τῆς [στή]λης ·" χαῖρε" λέγει [π]αρόδοις.

B.

Ἔτους θτ᾽, μη(νὸς) Λώου δ᾽.
Ἡ θήκη ἠγοράσθη ὑπὸ
Μάρ(κου) Αὐρ(ηλίου) Ἀττάλου
Τατᾶ, εἰς ἣν ταφήσε
τε αὐτός.

If the era be that of *Sulla*, then the inscription dates from the year 225 A.D.; if the era be that of *Cibyra*, then the inscription dates from 334 A.D. The former is most probably the true date.

In *A*, line 4, ΜΙΓΑ is the adverb *with*.

No. 19.

Makuf. In the wall of the Acropolis; the stone is very rough and was never polished. Bulletin de Correspondance Hellénique, 1885, *p.* 340.

```
▨▨▨▨IHΓOPACΘH▨▨▨▨ƆAYP·AIⱤⱯ
   ᴇ Λ▨ΔΙΟΝΥCΙΟΥ
NHCHNTINAⱰNHCATOYΠOMAΫPΠOCI
ΔΙΠΠOYᴇNHᴇNTAΦHCAIᴛᴇHAΓPIΠ
ΠINAKᴇONANΠᴇPIOYCⱮBOYΛHΘH
```

['Η θήκη] ἠγοράσθη [ὑπὸ] Αὐρ(ηλίας) Ἀ[γ]ρ[ιππί-]
νης, [Διονυσίου] ἥντινα? ὠνήσατο ὑπὸ Μ. Αὐρ. Ποσι-
δίππου, ἐν ᾗ ἐνταφήσαιτε ἡ Ἀγριπ-
πῖνα κὲ ὃν ἂν περιοῦσ[α] βουληθ[ῇ].

Between lines 1 and 2 the real names of the woman and her father have been inserted as an afterthought. The name of the woman may be Ἑλένη, Μελιτινή, Μελτίνη, or Μελίτιον, all of which are common. The reading of the inscription is certain.

No. 20.

Makuf.   Fragment in the wall near the Stadion.

```
▨▨▨▨▨▨▨▨▨▨▨▨▨▨▨▨▨▨
▨▨▨Λ Ι Τ Ι Ν Η C ⸔ Δ▨▨▨▨
▨▨▨O Ν Α Δ Ε Λ Φ Ο Ν Α Ι▨▨
▨▨▨Ν Τ Α Φ Η Ν Α Ι Ε Π Ι Α Π Ο
▨▨▨
```

. . . .

. . . . [Μελ]ιτινῆς Δ[ιονυσίου]?

. . . . [τ]ὸν ἀδελφὸν α[ὐτῆς]

. . . . [ἐ]νταφῆναι ἐπὶ Ἀπο-

[λλωνίου].

No. 21.

Makuf.   Unpolished stone serving as a post to a gateway.

```
Η Θ Η Κ Η Ε C Τ Ι Η Ν Ε Ι Κ Ι Ο Υ Τ Ο Υ Μ Ε Λ▨
▨Ι Ν Η C Ε Ν Η Τ Ε Θ Η C Ε Τ Α Ι Α Υ Τ Ο C
Ν Ε Ι Κ Ι Α C Ε Τ Ε Ρ Ο C Δ Ε Ο Υ Δ Ε Ι C
▨Π Ε Ι Ο Ι▨▨▨▨Ι Α C Α Μ Ε Ν Ο C Α Π Ο
▨Ε Ι C Ε Ι▨▨Κ Υ Ρ Ι Α Κ Ω Φ Ι C Κ Ω
            * Φ
```

Ἡ θήκη ἐστὶ[ν] Νεικίου τοῦ Μελ[ι-]

[τ]ινῆς, ἐν ᾗ τεθήσεται αὐτὸς

Νεικίας· ἕτερος δὲ οὐδεὶς,

[ἐ]πεὶ ὁ [ἐπεισβ]ιασάμενος ἀπο-

[τ]είσει [τ]ῷ κυριακῷ φίσκῳ

            (δηνάρια πεντακόσια).

Descent is very rarely reckoned μητρόθεν. It may be doubted whether it be the case here or not. At any rate we know from No. 15 that women held office in Heraclea, and it may be that Melitine was one of these distinguished women from whom it was an honor to reckon descent. It seems improbable that the town Melitene in Eastern Cappadocia is referred to in this inscription.

For ἐπεισβιάμενος, see *C.I.G.* 3996, 4007, 4190, 4360 *n*, etc.

No. 22.

Makuf. Stone forming a step in the doorway of a house. Elegiac distich.

```
Θ Α Υ Ε Μ Ε Τ Η Ν Ν Ε Ι Κ Η Ν Ξ Ε Ν Ι
Ο Ξ Ξ Ε Ν Ε Χ Α Ι Ρ Ε Π Ο Ρ Ε Υ Τ Ο Υ Μ Η
Δ Ε Ν Μ Ε Μ Υ Α Μ Ε Ν Ο Ξ Τ Η Ν
Ι Δ Ι Α Ν Α Λ Ο Χ Ο Ν
```

Θάψε με, τὴν Νείκην, Ξένιος· ξένε χαῖρε πορευτοῦ μηδὲν μεμψάμενος τὴν ἰδίαν ἄλοχον.

The reading ΠΟΡΕΥΤΟΥ is certain. If πορευτὸν or πορευτὸς can be made to mean *journey*, then the sense might be : *Xenios buried me, Nike. Prosper, stranger, in thy journey, and think kindly of thine own wife.*

Nos. 23-24.

Makuf. C.I.G. 3953 *b, from a copy of Schönborn.*

The inscriptions must read :

A.

['Η βουλ]ὴ καὶ ὁ δῆμος
Γλύκωνα Σωσ[θέ]νους κτί-
στην κ[αὶ] εὐε[ργέ]την τῆ[s]
π[όλ]εως
[κ]ατ[ὰ τὴν τ]οῦ Γλύκωνος
διαθήκην.

B.

Ἡ βουλὴ καὶ ὁ δῆμος
Μελίτιον Χαρ[μί]δου, γυναῖκα
Γλύκωνος Σω[σθ]ένους
κατὰ τὴν το[ῦ] Γλύκωνος
διαθήκην.

The above inscriptions are published (badly) in the *Bulletin de Correspondance Hellénique*, 1886, p. 519, as coming from Acharkieui near Tralleis.

May 30. Makuf to Davas, 4 h. 30 m. From Heraclea I journeyed to the southeast and south through the plain now known as Davas Ova, passing Tilkilly and Karakieui, and made as good a survey of the district as I could. I visited Tabae, now Davas, but found no inscriptions, and but few traces of an antiquity other than Turkish. Tabae is situated on a high hill in a gorge between two mountains, and is surrounded by cañons three or four hundred feet deep on all sides except one. On this one side it is approached by a bridge, which crosses a chasm where it is least deep; after the bridge is crossed a narrow neck of land, just wide enough for a roomy road, leads by a tortuous and laborious ascent to the town on the hill. When seen from any point in the plain, Davas seems to be situated on a neck or saddle of the mountains, and one is extremely surprised at the real topography of the place.

May 31. Davas to Medet, 4 h. 45 m. The plain of Tabae is one of extraordinary fertility; in antiquity it supported three cities, Heraclea, Tabae, and a third at Medet, to which Messrs. Paris and Holleaux give the name of Apollonia (see *Bulletin de Correspondance Hellénique*, 1885, p. 342 sqq.). The inscriptions published by these gentlemen (*Bulletin de Correspondance Hellénique*, 1885, p. 344 sqq.) were copied by me also (see my *Preliminary Report*, p. 6).

That Apollonia was a town of considerable importance and wealth is clear from a very substantial antique substructure in huge hewn stones of blue limestone. Upon these foundations there now stands the Mosque, which has without doubt superseded a prouder structure in honor of a pagan god. The neighboring hill, which certainly served as the Acropolis, contains no traces of antiquity, except the many architectural fragments of great weight and size which are found in the cemetery.

June 1. Medet *via* Kizildje to Kizildje Beïlik, 6 h. 3 m. The distance from Medet to Kizildje is 2 h. 20 m. Immediately east of Uzun Puñar we begin the ascent of the mountain which lies between Medet and Kizildje. The following inscription found at Kizildje identifies the site as that of Sebastopolis.

No. 25.

Kizildje [*Sebastopolis*]. *In the wall of the Djami. Length,* 1.25 *m.; width,* 0.50 *m. See my* Preliminary Report, *p.* 6.[1]

```
   ΑΥΤΟΚΡΑΤΟΡΙΝΕΡΒΑΤΡΑΙΑΝΩ
   ΑΡΙCΤΩΚΑΙCΑΡΙCΕΒΑCΤΩΓΕΡΜΑΝΙ
   ΚΩΔΑΚΙΚΩΠΑΡΘΙΚΩ
   ΚΑΙΤΩΔΗΜΩΤΩ̣CΕΒΑΤΟΠΟ
 5 ΛΕΙΤΩΝ·Π·CΤΑΤΙΟCΕΡΜΑCΑΓΟ
   ΡΑΝΟΜΗCΑCΚΑΙΠΑΡΑΦΥΛΑΖΑC
   ΚΑΙΤΕΙΜΗΘΕΙCΕΤΙΤΕΥΠΕΡ
   ΓΗCCΤΡΩCΕΩCΤΗCCΖ̄Ζ̄Ε
   ΔΡΑCΤΗCΕΝΤΩΤΕΤΡΑCΤΥ
10 ΛΩΤΟΥΓΥΜΝΑCΙΟΥΤΕΙΜΑΙC
   ΕΙΡΗΝΑΡΧΙΚΑΙCΠΑΛΙΝΔΕΥ
   ΠΕΡΑΝΑCΤΑCΕΩCΤΗCΝΕΙ
   ΛΗCΕΚΤΩΝΙΔΙΩΝΤΕΙΜΗΘΕΙC
   ΤΕΙΜΑΙCΔΙΑΝΥΚΤΟCCΤΡΑΤΗ
15 ΓΙΚΑΙCΚΑΙΑΠΟΔΟΧΕΥCΓΕ
   ΝΑΜΕΝΟCΓΧΜΑCΚΑΙΑΡΓΥ
   ΡΟΤΑΜΙΑC✱ΔΚΑΘΩCΚΑΙ
   ΔΙΑΤΩΝΥΗΦΙCΜΑΤΩΝ
        ΠΕΡΙΕΧΕΙ
```

Αὐτοκράτορι Νέρβᾳ Τραϊανῷ,
Ἀρίστῳ Καίσαρι Σεβαστῷ Γερμανι-
κῷ Δακικῷ Παρθικῷ

---

[1] Shortly after the appearance of my *Preliminary Report* the inscription was also published in the *Bulletin de Correspondance Hellénique*, 1885, pp. 346–347.

καὶ τῷ δήμῳ τῷ Σεβαστοπο-
5 λειτῶν Π. Στάτιος Ἑρμᾶς ἀγο-
ρανομήσας καὶ παραφυλάξας
καὶ τειμηθεὶς ἔτι τε ὑπὲρ
τῆς στρώσεως τῆς ἐ⟨ξ⟩ξέ-
δρας τῆς ἐν τῷ τετραστύ-
10 λῳ τοῦ γυμνασίου τειμαῖς
εἰρηναρχικαῖς, πάλιν δὲ ὑ-
[π]ὲρ ἀναστάσεως τῆς Νεί-
[κ]ης ἐκ τῶν ἰδίων τειμηθεὶς
τειμαῖς διὰ νυκτὸς στρατη-
15 γικαῖς καὶ ἀποδοχεὺς γε-
νάμενος γ′ (✳) μας′ καὶ ἀργυ-
ροταμίας ✳,δ καθὼς καὶ
διὰ τῶν ψηφισμάτων
19   περιέχει.

Lines 15–16, ΓΕΝΑΜΕΝΟΣ is certain; for this form see Ahrens,
Dial. II. p. 305; Kühner, *Ausführliche Grammatik*, I. p. 568,
Anmerk. 4; and *Mittheilungen des Deutschen Archaeologischen Instituts*, 1881, p. 258.   Line 16, X is probably ✳, the sign for Denars;
the number of the Denars is σμα′ (=241).   In line 17 I have placed
the stroke below the ͵δ, making it 4000, inasmuch as δ′ (4) seems too
small.

No. 26.

*Kizildje.   Stone serving as a step in the stairway of a
house.*

    ΣΤΟΥΤΕΙΜΟ
    Ω – ΜΙΘΡΗΝΖΩΖ
    ΡΑΤΙΑΕΤΕΡΩΔΕΟ
    ΝΚΗΔΕΥΣΑΙΤΟΛ
    Υ✳ΒΦΚΑΙΥΠΕΥ
    ΡΑΦΟΝΑΠΕΤΕ
    ΚΘΜΒ

. . . . τοῦ Τειμο[θέου]

. . . . Μίθρην Ζω[σίμου?]

. . . . . Σ]τρατία · ἑτέρῳ δὲ ο[ὐδενὶ]

ἔξεστι]ν κηδεῦσαι το . . . .

. . . . * ‚βφ' καὶ ὑπεύ[θυνος . . . .].

τῆς ἐπιγραφῆς ἀντίγ]ραφον ἀπετέ[θη εἰς τὰ

ἀρχεῖα, ἔτους] κθ', μη(νὸς) β'.

The inscription dates from the year 55 B.C., which corresponds
with the year 29 of the era of Sulla, or 52 A.D., which corresponds
with the year 29 of the era of Cibyra.

From Kizildje we return to the neighborhood of Medet, and thence
to Kizildje Beilik, the time occupied between Kizildje and Kizildje
Beilik being 3 h. 43 m.

June 2. Kizildje Beilik to Kizil Hissar, 5 h. 43 m. We follow
up the Harpasus River, and in 4 h. 9 m. the watershed is reached in
the neighborhood of Sarai Ova. A descent of 20 m. brings us to the
extreme northwestern corner of the Karayuk Bazar Ova. From this
point there is a gentle descent until within a short distance of Kizil
Hissar, which is situated at the western edge of the plain on the
slopes of the mountain. The distance from the watershed near Sarai
Ova to Kizil Hissar is 1 h. 34 m. I was detained a day at Kizil
Hissar, waiting for Messrs. Ramsay and Smith, who found it impossible
to be punctual.

June 4. Kizil Hissar, *via* Yüreghil, Yataghan, Kuyudjak, Kaïbazar,
Avshar, to Güine, 6 h. 34 m. One hour's travel brings us to the
easternmost limit of the plain. Henceforward the country is rolling,
the above-mentioned villages lying in narrow valleys, each with a
little stream of water. Kaïbazar is a large and prosperous village.
Güine is situated at the northern edge of a little valley, that is hemmed
in on all sides by mountains, except that the water finds an exit
through a narrow gorge to the west.

June 5. Güine, *via* Dodru Agha, Yazir, Gümüsh, Gumavshar, to
Tchamkieui, 6 h. 33 m. Leaving Güine we traverse the little valley
mentioned above, and cross a mountain ridge to a narrow valley
which leads west to Derekieui. Here are the ruins of an ancient
town. We continue to head nearly southeast, crossing a mountain

ridge to Dodru Agha. The distance from Güine to Dodru Agha is 3 h. 33 m.

No. 27.

Dodru Agha. In the wall of the Djami; length, 0.42 *m.; height,* 0.30 *m. Impression.*[1]

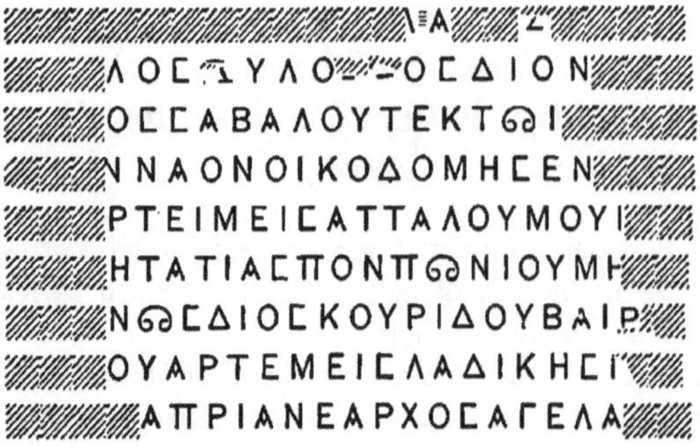
ΛΟϹ ΥΛΟ ΟϹΔΙΟΝ
ΟϹϹΑΒΑΛΟΥΤΕΚΤω Ι
ΝΝΑΟΝΟΙΚΟΔΟΜΗϹΕΝ
ΡΤΕΙΜΕΙϹΑΤΤΑΛΟΥΜΟΥΙ
ΗΤΑΤΙΑϹΠΟΝΠωΝΙΟΥΜ
ΝωϹΔΙΟϹΚΟΥΡΙΔΟΥΒΑΙ
ΟΥΑΡΤΕΜΕΙϹΛΑΔΙΚΗϹ
ΑΠΡΙΑΝΕΑΡΧΟϹΑΓΕΛΑ

. . . .

[Ἄττα]λος [Σ]υλο[σῶντ]ος? Διον[υσίου] . . .
. . . . . ος Σαβάλου τέκτω[ν] . . . . . .
. . . . [τὸ]ν ναὸν οἰκοδόμησεν . . . . . .
. . . . [Ἀ]ρτειμεῖς Ἀττάλου Μου[νδίωνος].
. . . . η Τατίας Πονπωνίου Μ[ήνιδος].
[Μί]νως Διοσκουρίδου Β[α]ι[βαίου?].
. . . . ου Ἀρτεμεῖς Λαδίκης . . . . .
. . . . [Κ?]απρία· Νέαρχος Ἀγελά[ου].

Line 6. If Βαβαίου be a correct conjecture, it must be the ethnic for the town Βαῖβαι in Caria.

Line 7. Λαδίκη stands for Λαοδίκη. The form Λαδίκη occurs in *C.I.G.* 3371, and in *Hdt.* 2, 181. Λαδικίη occurs in Dumont's *Inscriptions et Monuments Figurés de la Thrace*, p. 24, No. 53.

---

[1] Ligatures occur: line 3, ΜΗϹ; line 4, ΜΕ; line 5, ΝΠ; line 7, ΜΕ, ΗϹ.

No. 28.

Dodru Agha. In the wall of the Djami. Length, 0.40 m.; height, 0.40 m. C.I.G. 4380 s.

OCANTOYTOMNH
MEIONΛΔIKHCEIΘEΩN
KEXΩΛΩMENΩNTYXOI
TONΠICIΔIKΩN

'Ος ἂν τοῦτο τὸ μνη-
μεῖον ἀδικήσει θεῶν
κεχωλωμένων τύχοι-
τον Πισιδικῶν.

In *C.I.G.* 4380 *r* and *s* Franz gets rid of TYXOITON very unmethodically in *two* different ways. It is probably Pisidian Syntax for τύχοιεν (see Kühner, *Ausführliche Grammatik*, II, p. 18). At the date of this inscription the Dual was obsolete. "If any one violates the tomb, let *them* suffer for it at the hands of the Gods."

The two inscriptions which follow present two more examples of τύχοιτον, whatever it is.

No. 29.

Dodru Agha. In a field. Copied by W. M. Ramsay.[1]

EITICTOYTOTON
MNHMEIONAΔIKHCEI
ΘEΩNΠICIΔΩNKE
XOΛΘMENΘNTYX
OITON

Εἴ τις τοῦτο τὸ ⟨μ⟩
μνημεῖον ἀδικήσει
θεῶν Πισιδῶν κε-
χολωμένων τύχ-
οιτον.

---

[1] Ligatures occur: line 2, MNHME, HC; line 3, NΠ, NK; line 4, ME.

No. 30.

Dodru Agha. In the wall of the Djami.[1]

```
EITICTOYTOTO
MNEIMEIONAΔIKEI
ΘEⓌNΠICIΔIKΩNKE
XOΛⓌMENⓌNTYX
ⱣITON
```

Εἴ τις τοῦτο τό
μνειμεῖον ἀδικεῖ
θεῶν Πισιδικῶν κε-
χολωμένων τύχ-
οιτον.

No. 31.

*Dodru Agha. In the wall of the Djami. Long, 0.58 m.;
high, 0.30 m. Impression. Ç.I.G. 4380 t.*

```
MHΘICKAKOYP
CHCITOMNHMI
ONEIΔETICKAKOY
PΓHCIHTⲰENO
KOCHΛIⲰCEΛH
NH
```

Μηθὶς κακουρ-
[γ]ήσι τὸ μνημῖ-
ον, εἰ δέ τις κακου-
ργήσι ἤτω ἔνο-
κος Ἡλίῳ Σελή-
νῃ.

---

[1] Ligatures occur: line 2, MN; line 3, NTT, NK.

Concerning the late form ἥτω, see Kühner, *Ausführl. Gram.*, I. p. 666, 3.

About midway between Dodru Agha, on and around a small hill, there are sarcophagi and other traces of an ancient town. In the mosque of Yazir there are many ancient stones; sarcophagi are abundant, and I noticed the capital of a column belonging to the Christian period.

From Dodru Agha to Tchamkieui the time is three hours. Leaving Gümavshar we cross a low brushy barren hill to Tchamkieui. Here I met Messrs. Ramsay and Smith, who have explored the Karayuk Bazar Ova. During our separation of two days they had found the following six inscriptions.

No. 32.

Karayuk Bazar. Millarium forming part of a fountain outside of the town. Diameter, 21 inches. Copied by A. H. Smith and W. M. Ramsay.

```
    OIC▨▨OICHMWN
    AVTOKPATOICIN
    ΔIOKΛHTIANWKAI
    KAIM▨▨MIANWCEBB
 5  KAIKWCTANTIW
    KAIMAZIMIANW
    EΠIΦ'ΦKECAPCIN
```

Ṁ A

. . . . [τ-]

οἶς [κυρί]οις ἡμῶν
αὐτοκράτο[ρ]σιν
Διοκλητιανῷ (καὶ)

καὶ Μ[αξι]μιανῷ Σεβ(αστοῖς)
5 καὶ Κωσταντίῳ
καὶ Μαξιμιανῷ
ἐπιφ(ανεστάτοις) Κέσαρσιν·
Μί(λιον ἕν).

This is the first milestone from Themissonion, which was situated at Kara Eyuk Bazar.

No. 33.

Karayuk Bazar.   In the Djami.   Copied by W. M. Ramsay and A. H. Smith.

///////////MOYΛΠΙΟΣ
NWNOCYIONKYPINA
TPYΦWNAMEΓANANTΟ
NIANONAPXIIEPEATHCΚ
5 CIACXEIΛIAPXHCANTΚ
KAIΓENOMENONEΐΑ
XONCΠEIPHCΠPWTHC
OYΛΠIACΓAΛATWNEN
ΠACINΠPWTONTHCΠO
10 ΛEWCTEKAITHCEΠAPXE
ΛCTONEYEPΓETHNTHC
ΠATPIΔOC·HBOYΛHKAIΟ
ΔHMOC T HNANACTA
CINΠOIHCAMENHC
15 ANTWNIACAPICTHCAΛ
BIΛΛHCTHCEΓΓONHCAYToY
⚘ EKTWNIΔIWN ⚘

. . . . Μ. Οὔλπιο(ν),
['Αγ]νωνος? υἱὸν, Κυρίνα,
Τρύφωνα μέγαν 'Αντ[ω]-
νιανὸν ἀρχιερέα τῆς ['Α]-
5 σίας χειλιαρχήσαντ[α]
καὶ γενόμενον ἔ[π]α[ρ]-
χον σπείρης πρώτης
Οὐλπίας Γαλατῶν, ἐν
πᾶσιν πρῶτον τῆς πό-
10 λεώς τε καὶ τῆς ἐπαρχε[ί-]
[α]ς, τὸν εὐεργέτην τῆς
πατρίδος, ἡ βουλὴ καὶ ὁ
δῆμος· τὴν ἀνάστα-
σιν ποιησαμένης
15 'Αντωνίας 'Αρίστης 'Αλ-
βίλλης τῆς ἐγγόνης αὐτοῦ
ἐκ τῶν ἰδίων.

No. 34.

Karayuk Bazar. Copied by W. M. Ramsay and A. H. Smith.[1]

ΑΙΜΟΥΝΑΝΙΣΑΠΟΛ
ΛΩΔΟΣΠΛΕΥΡΟΥ
ΜΑΝΗΔΙΚΑΙΧΟϜΔΑΔΗ

Αἴμου Νανὶς 'Απόλ-
λωδος Πλεύρου
Μάνηδι καὶ Χο[ρ]δάδῃ.

The inscription is puzzling, and the last four names are new and strange.

---

[1] Ligature of NH in line 3.

No. 35.

*Tchamkieui. Drum of a red column. Copied by
W. M. Ramsay.*

```
   T O I Ϲ Θ E ๛ N
   A Y T O K P A T O
   Ϲ E Π T I M I ๛ Ϲ E Y H
   T I N A K I Ϲ E B A Ϲ
 5 Λ Δ░░░░░H N I K ๛ Π
   K A I Λ░░░░O K P A T O
   A Y P H Λ I ๛ A N T ๛ N
   Ϲ E B A Ϲ T ๛░░░░░░░
   ░░░░░░░░░░░░░░░░░
10    Γ Λ Δ ๛ N I A
   Ϲ E P A Ϲ T H ı ı ı-ı ʇ Γ ı ʌ ⱽ Λ
      Λ I : O ʜ B Y ı
```

Τοῖς θεῶν [ἐπιφανεστάτοις]
Αὐτοκράτο[ρι Καίσαρι Λουκίῳ]
Σεπτιμίῳ Σευή[ρῳ Εὐσεβεῖ Περ-]
τίνακι Σεβασ[τῷ Ἀραβικῷ]
5 [Ἀ]δ[ιαβ]ηνικ[ῷ] Π[αρθικῷ Μεγίστῳ]
καὶ [Αὐτ]οκράτο[ρι Καίσαρι Μάρκῳ]
Αὐρηλί[ῳ] Ἀντων[είνῳ Εὐσεβεῖ]
Σεβαστῷ [καὶ Ποπλίῳ Σεπτιμίῳ]
[Γέτᾳ ἐπιφανεστάτῳ Καίσαρι]
10 [καὶ Ἰουλ]ί[ᾳ] Δ[όμν]ᾳ
Σεβαστῇ [μ]η[τρὶ κάστρων]·
['Απ]ὸ [Κι]βύ[ρας]

No. 36.

Usuftcha. Circular basis beside the entrance to the Djami. Copied by A. H. Smith.

ΟΔΗΜΟΣΚΑΙΟΙΠΡΑΓΜΑ
ΤΕΥΟΜΕΝΥΕΝΤΑΥΘΑΡΟΜ▨
ΟΙΕΓΙΜΗΣΑΝΜΙΘΡΗΝΕΥΡ▨
ΧΡΥΣΩΙΣΤΕΦΑΝΩΙΤΙΜΟΙΣ▨Δ▨
ΚΑΙΕΙΚΟΙ

'Ο δῆμος καὶ οἱ πραγμα-
τευόμενυ ἐνταῦθα Ρομ[αῖ-]
οι ἐ[τ]ίμησαν Μίθρην Εὐ[βίου?]
χρυσῶι στεφάνωι τιμ(ί)ωι . . . .
καὶ εἰκό[νι γραπτῇ?]

For an enumeration of the places where Roman merchants were settled, see *Papers of the American School of Classical Studies at Athens*, I. p. 31, and III. p. 339, No. 473.

No. 37.

Aghlan. Site of old city near Aghlan; on a red column with capital. Copied by W. M. Ramsay, who makes a note that every symbol in the last line is certain except Λ, which is probably Α.

ΜΗΝΙΣ▨ΠΟΛΩΝ
ΟΥΕΑΥΤΩΖΩΝ
ΚΑΙΝΑΝΑΤΗΓΥΝΑΙ
ΖΩСΗ   ΚΙ
ΙΕΡΕΥСΔΗΜΗΤΡΟС
ΚΛΙСΑΟΑΖΟΥ

Μῆνις ['A]πολων[ί-]
ου ἑαυτῷ ζῶν
καὶ Νάνᾳ τῇ γυναικὶ
ζώσῃ
ἱερεὺς Δήμητρος
κ[α]ὶ Σαοάζου.

June 6. Tchamkieui to Derekieui, 2 h. 54 m. We retrace our steps to the neighborhood of Gümavshar, and thence to Derekieui, passing around a high mountain on our right. At Yaghmur Tash (= stone rain) the plain is strewn with many huge bowlders, having been precipitated down from the almost perpendicular heights. The plain here really looks as though it had rained stones. Half an hour north of Derekieui in the plain there are substantial foundations, possibly of a temple. On the top of the mountain immediately east of Derekieui the villagers report a Kale and inscriptions, but having promised to meet Messrs. Ramsay and Smith at a fixed time, it was not possible for me to investigate the site.

June 7. Derekieui to Karamanlü, 8 h. 36 m. Leaving Derekieui we went up the narrow valley and passed the ruins mentioned on June 5. Thence we recross the mountain ridge to Dodru Agha. Leaving Dodru Agha, 33 m. travel east over a rough country brings us to the foot of the high, rugged, and in places almost impassable Eshlèr Dagh. After a climb of 3 h. 50 m. the final summit of the mountain is reached. A descent of 1 h. 39 m. brings us to the foot of the Eshler Dagh, at the westernmost limit of the plain of Karamanlü. Thence, passing Gultchan, we reach Karamanlü in 1 h. 26 m.

Nos. 38–40.

Karamanlü. Quadrangular cippus at the fountain west of the village. Height, 1.45 *m.; high between the mouldings,* 1.05 *m.; width,* 0.50 *m.* Bulletin de Correspondance Hellénique, 1878, *p.* 246 *sqq. Impressions.*

A.[1]

KAΛΠOPNIOCΔAOC
COΨPNOΨIEPACAMENOC

AYPHPAKΛEIΔHCNEAPXOYI
EPACAMENOC ˙
5   AYPNEIKOΛAOCATTH
ΔICTOYOCAEIIEPACATO
⦰YPΔHMHCXAPHTOCΓNAI
OYIEPACATO
ΔHMHCNEIKAΔA
10   KAΔAOΨIEPACAMENO⦰
  OC   MHNICTPIᶜΛATYΠO⦰
  NOC IEPACATO
  OC   EITAΛIKOCKACIOY
    ΠACAIEPACATO
15   ATTACAΠOΛΛWNIO
  OC   YTPICTOYIEPEOCIEPA

Var. Lect.

2 init. The *Bulletin* reads OYINOY.

3. The *Bulletin* omits I in fine.

4.   "   "   omits E in init.

6.   "   "   reads Δ in init.

7.   "   "   reads A in init. OCΓNAI in fine.

10.   "   "   inserts in the line the OC, which is on the edge of the stone and belong to inscription *B*.

11. The *Bulletin* inserts NOC, which belongs to *B*, as above.

12.   "   "   omits the OC belonging to *B*, and reads E in init.

14.   "   "   omits Π in init., and indicates the loss of two letters.

16.   "   "   inserts in the line in small letters the OC on the edge of the stone, and which belongs to *B*.

---

[1] The small letters to the left of the uncial text belong to inscription *B*, which occupies the side of the stone to the left of inscription *A*. In line 1, Y has been omitted in the name. In line 5, the stone is uncut after ATTH. In line 11, the C in TPICΛA is very small. Lines 1 and 2 are on the moulding.

```
NOC          C A T O
             A Y P K A Λ Λ W N A Π O Λ Λ W N I O▨
             Γ I E P E O C I E P A C A M E N O C
20           A Y P T P O Π I M O C M O
NOC          K W T O Y E I E P A C A
             T O
```

Var. Lect.

17. The *Bulletin* inserts in the line in small letters ΦΙΟC instead of the NOC on the edge, and which belongs to *B*.
18. The *Bulletin* reads a small C above the line in fine.
21.   "   "   inserts the NOC belonging to side *B*.
22. The TO of this line completes inscription *A*. The additional lines 23-27 given by the *Bulletin* belong to side *B*, and are directly opposite the lines indicated in my uncial text of inscription *B*.

B.[1]

[Twenty-one lines so badly defaced as to be hopeless, although single letters at the beginning of the lines are distinct.]

```
22   A Π O Λ Λ W▨▨▨▨▨▨▨▨▨▨▨▨▨▨▨▨▨▨
     K A C I O▨▨▨▨▨▨▨▨▨▨▨▨▨▨▨▨▨▨▨
     ⊐ E K O Y ⊐▨▨ O▨▨▨▨▨▨Λ C A M E N O ⊐
```

Var. Lect.

22. The *Bulletin* reads AΠ alone.
23.   "   "   reads E in fine.
24.   "   "   reads ΓE . O alone.

---

[1] The letters to the right of the uncial text belong to this side as indicated, but they are inscribed on the side of the stone to the right of inscription *B* (see *Bulletin de Correspondance Hellénique*, 1878, p. 246, lines 23-27, and p. 249, lines 23-27). In line 31, EIA, by error of the stonecutter is certain. In line 3, the reading ΛΟΥΧΝΕΙ is certain, the X being probably a numeral.

25 ΝΕΙΚΑΔΑϹΜΗΝΙΔΟϹΚΑΔΑΥΟΥΙΕΡΑϹΑΜΕ
ΜΕΝΕϹΘΕΥϹ▨ΩϹΙΜΟΥΙΕΡΑϹΑΜΕ
ΔΗΜΗΤΡΙΟΞΜΑΡΚΕΛΛΟϹΙΕΡΑϹΑΜΕΝΟϹ
ΚΑΡΝΟϹΑΤΤΟΛΛΩΝΙΟΥΙΕΡΕΟϹΙΕΡΑϹΑΜΕΝΟ
ΝΕΑΡΧΟϹΑΤΟΛΛΩΝΙΟΥΜΟΥΝΔΙΩ   ΝΟϹΙΕΡΑϹΑΜΕΝΟϹ
30 ΚΙΔΡΑΜΟΑϹΔΙϹΝΕΟϹΙΕΡΑϹΑΜΕΝΟϹ
ΚΑϹΤΩΡΜΟΛΥΚΟϹΕΙΑΡΑϹΑΜ   ΕΝΟϹ
ΑΥΑΤΤΟΛΛΩΝΙϹΙΕΡΕΟϹΝΕΟϹΙΕΡΑϹ   ΑΜΕΝΟϹ
ΝΕΙΚΟΛΑϹΑΤΤΑΛΟΥΧΝΕΙΚΟΛΑ▨ΥΙΕ
ΑΥΡΟΝΗϹΙΜΟϹΜΕΝΕϹΤΕΟϹ▨ΕΡΑϹΑΜΕ   ΝΟϹ
36 ΑΡΤΕΙΜΗϹΧΑΡΗΔΟϹΝ▨

Var. Lect.

25. The *Bulletin* reads . ΗΝ . . Ο and nothing more.
26. "   "   omits entirely.
27. "   "   reads Κ and nothing more.
28. "   "   omits entirely.
29. "   "   reads ΠΕΑ and nothing more.

30. The *Bulletin* omits the Α in ΟΑϹ.
31. "   "   reads ΚΑϹΤΩΡ . . . . . ΚΟϹ and nothing more.
32. The *Bulletin* ends the line with ΙΕΡΑ . . .
33. "   "   reads ΝΕΙΚΕΛΑϹ in init.
34. The *Bulletin* reads ϹΙΕ where I read Ϲ▨Ε, and ends the line with ΜΕ.
35. "   "   reads ΝΗϹ instead of ΜΗϹ; it reads ϹϹ▨ in fine.

TOYЄIЄPACAMЄNOC
AYPKACIOCTPICГ▨
OYIЄPACATO

▨▨▨▨▨▨▨▨▨▨▨▨▨▨▨▨▨▨▨

40   M O▨▨▨IЄPACATO

Var. Lect.

36. The *Bulletin* reads TOYIЄ in init.
37.   "   "   reads TPICIᴌ▨ in fine.
39.   "   "   does not indicate the defaced line.
40.   "   "   reads . . . . MOY . . ЄPAC.

C.[1]

MHNICTPICOCAЄIIЄPAC
AMЄNOC

AYPATTAᴌOCOCAЄIKᴧHPONON▨▨
                              MЄNO▨
AYPOCAЄICATTAᴌOYNЄOCIЄPAC▨
5 ▨AYNЄIKAΔACMHNIΔOCKAΔAYOYIЄPACAT
▨AYPHATTAᴌOCNЄIKOᴧAOYIЄPACAN▨
▨AYPHᴧIOCMHNIᵒATTAᴌOY   ЄNO▨
KAᴧᴧIKᴧЄOYCIЄPACAMЄNOC
▨ᴧAPKЄᴧᴧOCNЄOCIЄPACAMЄNOC;

Var. Lect.

3. The *Bulletin* reads M in fine, and does not indicate a break.
4.   "   "   ends the line IЄPAC . . . . . ., and does give the MЄNO▨ above the line.
5. The *Bulletin* reads A in init., and closes the line with IЄP▨.
6.   "   "   reads A in init. and M▨ in fine, failing to give the ЄNO▨ below the line.
9. The *Bulletin* reads APKЄᴧAOC in init.

---

[1] Ligatures occur: line 5, HN; line 10, MHN. In line 7, the C between I and A is very small. Lines 1 and 2 are on the moulding.

10 ░ΑΤΤΑΛΟϹΜΗΝΙΔΟϹΙϹΚΑΛΟ░

ΧΑΡΗϹΑΠΟΛΛΩΝΙΟΥΙΕΡΕ
ΟϹΙΕΡΑϹΑΜΕΝΟϹ
ΡΟΥΦΕΙΝΟϹΙΕΡΑϹΑΜΕΝΟϹ
░ΙΑϹΚΟΥΡΙΔΗϹΔΙϹΕΥΤΥΚΟΥ
15 ░ΕΡΑϹΑΜΕΝΟϹ
░ΤΑΛΟϹΒΚΑΛΛΙΚΛΕΟΥϹΙΕΡΑϹΑΜΕΝΟϹ
░ΜΗΝΙϹΝΙΚΑΔΑΔΟϹΜΗΝΙΔΟϹΚΑΔ░
░ΥΙΕΡΑϹΑΜΕΝΟϹ
ΓΕΩΡΓΟϹΓΑΕΙΟΥΙΕΡΑ
20 ϹΑΜΕΝΟϹ

Var. Lect.

10. The *Bulletin* reads ΛΟ in fine, and after line 10 indicates a
defaced line ; it is a natural gap.
16. The *Bulletin* reads .. ΤΑΛΟϹΚΑ, omitting the Β between Ϲ
and Κ ; it closes the line with ΑΜ░.
17. The *Bulletin* reads ΜΗΝΙϹΝΙΚΑΔΟϹ in init., and ΚΑΛ░ in
fine.
19. The *Bulletin* reads . ΕΩ in init.

A.

Καλπόρνιος Δάος
Σούρνου ἱερασάμενος
Αὐρ. Ἡρακλείδης Νεάρχου ἱ-
ερασάμενος ·
5 Αὐρ. Νεικόλαος Ἄττη
[δ]ὶς τοῦ Ὀσαεὶ ἱεράσατο ·
[Α]ὐρ. Δημῆς Χάρητος Γναίου ἱεράσατο ·
Δημῆς Νεικάδα
10 Καδάου ἱερασάμενο[ς] ·
Μῆνις τρὶς λατύπο[ς]
ἱεράσατο ·

[Ε]ἰταλικὸς Κασίου
Πασᾶ ἱεράσατο·
15 *Ἄττας Ἀπολλωνίο-
υ τρὶς τοῦ Ἱερέος ἱερά-
σατο·
Αὐρ. Κάλλων Ἀπολλωνίο[υ]
γ' Ἱερέος ἱερασάμενος·
20 Αὐρ. Τρόπιμος Μο-
κωτοῦ εἱεράσα-
το.

B.

22 Ἀπολλώ[νιος . . . . ἱερασάμενος]·
Κάσιο[ς . . . . ἱερασάμενος]·
. . . εκο . . . . [ἱερα]σάμενος·
25 Νεικάδας Μήνιδος Καδαύου ἱερασάμε[νος]·
Μενεσθεὺς [Ζ]ωσίμου ἱερασάμε[νος]·
Δημήτριος Μάρκελλος ἱερασάμενος·
Κάρνος Ἀπολλωνίου Ἱερέος ἱερασάμενο[ς]·
Νέαρχος Ἀπολλωνίου Μουνδίωνος ἱερασάμενος·
30 Κιδραμόας δὶς νέος ἱερασάμενος·
Κάστωρ Μόλυκος εἱαρασάμενος·
Αὐ[ρ]. Ἀπολλῶνις Ἱερέος νέος ἱερα[σ]άμενος·
Νεικόλας Ἀττάλου [δ'?] Νεικολά[ο]υ ἱε[ρασάμενος]·
Αὐρ. Ὀνήσιμος Μενεστέος [ἱ]ερασάμενος·
35 Ἀρτειμῆς Χάρηδος [Μοκω-]
τοῦ εἱερασάμενος·
Αὐρ. Κάσιος τρὶς Γ[αί-]
ου ἱεράσατο·
[Ὁ δεῖνα τοῦ Ζωσί-]
40 μο[υ?] ἱεράσατο.

C.

Μῆνις τρὶς 'Οσαεὶ ἱερασ-
άμενος·
Αὐρ. "Ατταλος 'Οσαεὶ κληρονό[μος]·
Αὐρ. 'Οσαεὶς 'Αττάλου νέος ἱερασ[ά]μενο[ς]·
5 [Α]ὐ[ρ]. Νεικάδας Μήνιδος Καδαύου ἱεράσατ[ο].
[Α]ὐρή(λιος) "Ατταλος Νεικολάου ἱερασά[μ]ενο[ς]·
[Α]ὐρήλιος Μῆνις 'Αττάλου
Καλλικλέους ἱερασάμενος·
[Μ]άρκελλος νέος ἱερασάμενο[ς]·
10 "Ατταλος Μήνιδος 'Ισκάλου·
Χάρης 'Απολλωνίου 'Ιερέ-
ος ἱερασάμενος·
'Ρουφεῖνος ἱερασάμενος·
[Δ]ιασκουρίδης δὶς Εὐτύκου
15 [ἱ]ερασάμενος·
["Ατ]ταλος β' Καλλικλέους ἱερασάμενο[ς]·
[Μ]ῆνις Νεικάδαδος Μήνιδος Καδ[αύ-]
[ο]υ ἱερασάμενος
[Γ]εωργὸς Γαείου ἱερα-
20 σάμενος.

Nos. 41-42.

*Karamanli. Quadrangular cippus near the fountain, west of
the village. Height,* 1.20 *m.; within the mouldings,* 0.87 *m.;
width,* 0.44 *m.* Bulletin de Correspondance Hellénique,
1878, *p.* 250. *Impression.*[1]

---

[1] Lines 1-2 are on the top moulding. The l at the end of line 3 is very small.
The letters to the right of the uncial text belong to this inscription as indicated,
but are on the side of the stone immediately to the right of *A* (see *Bulletin de
Correspondance Hellénique,* 1878, p. 251, lines 12-14). In line 23, MO in ligature
probably stands for MOYNΔIWN.

A.

▨▨▨▨▨▨▨ΘΗΤΥΧΗΕΤΟΥϹ
Θ ▨▨▨▨▨▨▨▨▨▨Π.Ι.Λ.▨▨▨
ΥΠΕΡϹШΤΗΡΙΑϹΑΥΤШΝΚΑΙ
ϹШΤΗΡΙΑϹϹΕΒΗΡΟΥΚΑΙ
5 ΦΑΥϹΤΕΙΝΗϹΚΑΙΔΗΜΟΥ
ΟΡΜΗΛΕШΝΕΠΙΑΕΙΘΑΛ
ΟΥϹΠΡΑΓΜΑΤΕΥΤΟΥ
ΑΠΟΛΛШΝΙϹΑΤΤΑΛΟΥΜ
ΟΥΝΔΙШΝΟϹΠΡΟΑΓШΝ
10 ΝΕΑΡΧΟϹΑΠΟΛΛШΝΙΟΥ
ΑΠΟΛΛШΝΙϹΔΙϹΕΙΕΡΕΟϹΚ
ΟΥΡΠΑΕΡΜΑΙΟϹΔΗΜΗΤΡΙΟΥ
ΜΗΝΙϹΑΤΤΑΛΟΥΜΟΥΝΔΙШΝΟϹ
ΕΙΤΑΛΙΚΟϹΤΡΟΦΙΜΟΥ
15 ΑΤΤΑΛΟϹΑΠΟΛΛШΝΙΟΥ
ΧΑΡΗϹΜΗΝΙΔΟϹΚΟΥΜΑΔΙϹ
ΜΗΝΙϹΔΙϹΝΕΙΚΑΔΟΥ
ΜΗΝΙϹΔΙϹΟϹΑΕΙΚΑϹΤШΡ
ΝΕΙΚΑΔΑΔΟϹΜΗΝΙϹΧΑΡΗϟ
20 ΤΟϹϹΚΥΤΕΟϹΝΕΙΚΑΔΑϹ
ΜΗΝΙΔΟϹΝΕΙΚΑΔΟΥ
ΖШϹΙ [vacat] ΜΗΝΙϹΓΝΑΙΟΥ

Var. Lect.

1. The *Bulletin* reads Τ ΧΗΖΓΟΥ.
2.   "   "   reads Π̈.
3.   "   "   reads · Π in init. and ΑΙ in fine.
6.   "   "   reads Λ in fine.
7.   "   "   reads ΟΥᶜ in init. and ΤΟΥΜ in fine.
8.   "   "   omits entirely.
10. "   "   reads ΠΟΛ: ·
16. "   "   reads ΚΟΥΤΙΑΛΙ▨ in fine.
19. "   "   reads ΝΕΙΚΑΔΟϹ in init. and Η in fine.

ΧΑΡΗΤΟΣΜΝΕΙΚΑΔΑΣΤΡΙΣ

ΜΗΝΙΣΝΕΙΚΑΔΟΓΤΡΙΣ

25 ΜΗΝΙΣΝΕΙΚΑΔΟΥ

ΜΗΝΙΣΔΙΣΔΡΑΥΚΩΝ

ΚΑΛΛΙΚΛΗΣΔΙΣΕΡΜΑΙΣΔΙΣ

ΜΗΝΙ▨▨▨▨▨▨▨▨▨▨▨▨▨▨

▨▨▨▨▨▨▨▨▨▨▨▨▨▨▨▨▨▨

30 ▨▨▨▨▨▨▨▨▨▨▨▨▨▨▨▨▨▨

ΑΠΟΛΛΩΝΙΣΔΙΣΜΟΥΝΔΙΩ   ΝΟΣ

ΜΗΝΙΣΚΑΣΤΟΡΟΣ

ΑΠΟΛΛΩΝΙΣΤΡΙΣΕΙΕΡΕΟΣΚΟ   ΥΡΠΑ

ΣΤΡΑΤΩΝΚΩΒΕΛΛΕΩΣΤΕ   ΙΜΟΘΕΟΥ

Var. Lect.

23. The *Bulletin* declares the small O above the Μ a point.
24. " " does not indicate the break in the upper part of the Υ.
26. " " reads ΔΡΑΥΚΩΝ.
27. " " reads ΚΑΛΛΙΚΛΗ▨▨▨ and nothing more.
28. " " omits and says "quatre ligues martelées"; there are but two wholly defaced lines, 29 and 30.
31. The *Bulletin* read Ш▨ in fine, and does not discover that the ΝΟΣ around the corner is the end of the line.
33. The *Bulletin* reads ΣΚ▨ in fine, and does not discover that the ΥΡΠΑ around the corner is the end of the line.
34. The *Bulletin* reads ΣΤΡΑΤΩΝ'ШΒΕΛΛΑΖШΣΙ, and does not discover that the ΙΜΟΘΕΟΥ around the corner is the end of the line. After line 11 of inscription *B*, the *Bulletin* gives parts of the ends of lines 31, 33, 34 as belonging to inscription *B*, whereas in reality they belong to *A*, as indicated in my uncial text.

B.

On the second face of the same stone, but by a different stone-cutter. Bulletin de Correspondance Hellénique, 1878, *p.* 251.[1]

---

[1] None of the lines were ever carried clear across the stone. Line 3 was never finished, owing, no doubt, to the carelessness of the engraver.

```
   ΜΕΝΙΣΘΕΥΣΟΝΗΣΙΜΟΥ
   ΙΤΑΛΙΚΟΣΔΙΣΙΤΑΛΙΚΟΥ
   ΜΕΝΙΣΘΕΥΣΔΙΣΜΕΝΙΣ
   ░░ΝΑΙΟΣΜΕΝΙΣΘΕΟΣ
 5 ▨ΠΟΛΛΩΝΙΟΣΣΤΡΑΤΩΝΟ▨
   ΧΑΡΗΣΑΠΟΛΛ
   ΩΝΙΟΥΕΙΕΡΕΟΣ
   ΜΟΥΝΔΙΩΝΟΣ              •
   Κ ΑΣΤΩΡΜΗΝΙΔΟΣ
10 ΜΟΛΥΚΟΣ
         [blank]
   ΜΟΥΝΔΙΩΝΟΣ
```

Var. Lect.

1. The *Bulletin* gives all the Σ as C.
3.   "   "   reads M in init.
4.   "   "   reads . ΝΑΙΟΣ in init. and Σ in fine.
5.   "   "   reads Α in init. and ΤΡΑΤ.
6.   "   "   reads ω in fine.
8.   "   "   reads ΜΟΥ in init.
9.   "   "   reads Κ in init.
11.   "   "   reads Μ in init.

Lines 12–14 of the *Bulletin* are the ends of the lines 31, 33, 34 of inscription *A*.

A.

['Αγα]θῇ Τύχῃ · Ἔτους
θ . . [μηνὸς] . . ΠΙΛΙ . .
[ὑ]πὲρ σωτηρίας αὐτῶν κα[ὶ]
σωτηρίας Σεβήρου καὶ
5 Φαυστείνης καὶ δήμο[υ]
Ὀρμηλέων ἐπὶ 'Αειθαλ-
οῦς πραγματευτοῦ
'Απολλῶνις 'Αττάλου Μ-
ουνδίωνος προάγων ·

10 Νέαρχος Ἀπολλωνίου·
Ἀπολλῶνις δὶς Εἱερέος Κ-
ουρπᾶ· Ἑρμαῖος Δημητρίου·
Μῆνις Ἀττάλου Μουνδίωνος.
Εἰταλικὸς Τροφίμο[υ]·
15 ῎Ατταλος Ἀπολλωνίου·
Χάρης Μήνιδος Κουμᾶ δὶ[ς]·
Μῆνις δὶς Νεικάδου·
Μῆνις δὶς Ὀσαεὶ· Κάστωρ
Νεικάδαδος· Μῆνις Χάρ[η]-
20 τος Σκύτεος· Νεικάδας
Μήνιδος Νεικάδου,
ζῶσι· Μῆνις Γναίου
Χάρητος Μο(υνδίωνος)· Νεικάδας τρὶς·
Μῆνις Νεικάδο[υ] τρὶς·
25 Μῆνις Νεικάδου·
Μῆνις δὶς Δραύκων(ος?)·
Κ[α]λλι[κ]λῆς δὶς· Ἑρμ[α]ῖς δὶς
Μήνι[δος]· · · · · · · · · · ·
· · · · · · · · · · · ·
30 · · · · · · · · · · ·
Ἀπολλῶνις δὶς Μουνδίωνος·
Μῆνις Κάστορος·
Ἀπολλῶνις τρὶς Εἱερέος Κουρπᾶ·
Στράτων Κωβελλέως Τειμοθέου·

B.

Μενισθεὺς Ὀνησίμου·
Ἰταλικὸς δὶς Ἰταλικοῦ·
[Μ]ενισθεὺς δὶς Μενισ(θέος)·
[Γ]ναῖος Μενισθέος·
5 [Ἀ]πολλώνιος Στράτωνο[ς]·

Χάρης 'Απολλ-
ωνίου Εἱερέος
Μο[υ]νδίωνος ·
[Κ]άστωρ Μήνιδος ·
10 Μόλυκος
[Μ]ουνδίωνος.

No. 43.

*Karamanlii.   Quadrangular cippus at the fountain west of the
village.   Height,* 1.55 *m.; height within mouldings,* 1.03 *m.;
width,* 0.50 *m.*   Bulletin de Correspondance Hellénique,
1878, *p.* 253.   *Impression.*[1]

A I▨▨▨▨▨▨▨▨▨▨▨▨▨▨▨▨
T B ▨▨▨▨▨▨▨▨▨▨▨▨▨

C Ш T H P I A C A N I A C A Y P H
I A C Π A Y C T P N H C T H C K▨▨
5 I C T H C C Π I E Π I T P O Π O Y

Var. Lect.

1.   The *Bulletin* reads A.
2.   "   "   reads ṬIB.
3.   ·"   "   reads ШTH in init., and PH · in fine.
4.   "   "   reads IACΠAYCTPINHCTHCI▨
5.   "   "   reads EΠIEΠITPO▨▨ΠO\ in fine.

---

[1] Lines 1–9 are on the mouldings. Line 2 is certainly TB and not TIB. In
line 4, ΠAYCTPNHC is distinct. Line 6 has very distinctly ΠPΓ^AΛATЄY, a
serious blunder of the engraver. In line 9 end, ΠPA is certain, not ΠPOA. In
line 11 an omega is written upside down. In line 12, TPЄC for TPIC. In line 19
there is certainly but one C where there should be two. Iu line 22, TYΔPAHON
is distinct and certain, possibly an error for something like TYΔPAIШN. It is a
native name.

▨▨ΤΤ▨▨ΟⲤΚΕΠΡΓΛΑΛΑΤΕΥ
ΤΟΥΚΕΥΠΕΡⲤΩΤΗΡΙΆⲤ
ΤΟΥΔΗΜΟΥΟΡΜΗΛΕΩΝ

ΑΥΡΚΡΑΤΕΡΟⲤΚΛΑΥΔΙΟΥΠΡΑ
10   ΓΩΝΕⲤΤΗⲤΕΝΤΟΝΒΩΜΟΝ
ΕΚΤΥΝΕΙΔΙΩΝΑΝΑΛΩΜΑΤΩΝ
ΚΑⲤΙΟⲤΤΡΕⲤΠΑΝⲤΑ   ·
ΖΩⲤΙΜΟⲤΑΠΟΛΛΩΝΙΟ
ΥΔΑΡΝΟⲤΠΡΟΑΓΩΝ
15   ΤΡΟΠΙΜΟⲤΙΤΑΛΙΚΟΥ
▨ΑΛΛΙΚΛΗ▨ΙΤΑΛΙΚΟΥ

▨ΤΑΛΙΚΟⲤΚΛⲤΙΟΥ
▨ΑΙΟⲤΤΡΙⲤ
▨ΟΥΠΙΝΟⲤΩΚΡΑΤΟ▨
20   ▨ΛΕΞΑΝΔΡΟⲤΔΙⲤ
▨ΩⲤΙΜΟⲤΔΙΑⲤΚΟΥΡΙΔΟΥ

ΑΠΟΛΛΩΝΙΟⲤΤΥΔΡΑΗΟΝ
ΑΠΟΛΛΩΝΙΟΓΠΑΡΜΟΝ▨
▨ΤΑⲤΚΟΥΡΙΔΗⲤΔ″Ⲥ
25   ▨ΤΤΑΛΟⲤΓΕΡΜΕΟΥ
ΗΡΑΚΛΙΔΗⲤΝΕΑΡΚΟΥ

Var. Lect.

6. The *Bulletin* reads ΟⲤΚΕΠΡΓΜΑΤΕΥ.
11.   "   "   reads ΤΥΝ.
12.   "   "   reads · ΑⲤΙΟⲤΤΕⲤ in init.
13.   "   "   reads ΩⲤ in init., omitting Ζ.
14.   "   "   omits Υ in init.
17.   "   "   reads ΚΑⲤ.
20.   "   "   reads · ΙΕ in init.
23.   "   "   reads ΙΠ in init.
24.   "   "   reads · ΙΑ in init.

The date of the inscription is 218 A.D. Concerning πραγματευτής, see Lightfoot, *Apostolic Fathers*, Part II. Vol. I. p. 616, and *C.I.G.* 3101.

'Α[γαθῇ Τύχῃ · Ἔτους]
τβ', [μηνὸς . . . . · ὑπὲρ]
σωτηρίας 'Ανίας Αὐρη[λ-]
ίας Παυστ(ί)νης τῆς κ[ρατ-]
5 ·ίστης [ἐ]πὶ ἐπιτρόπο[υ]
. . . . . ος κὲ πρ[α]γ[μα]τευ-
τοῦ, κὲ ὑπὲρ σωτηρίας
τοῦ δήμου 'Ορμηλέων ·
Αὐρ. Κρατερὸς Κλαυδίου πρ(ο)ά-
10 γων ἔστησεν τὸν βωμὸν
ἐκ τῶν εἰδίων ἀναλωμάτων ·
[Κ]άσιος τρὲς Πάνσα ·
Ζώσιμος 'Απολλωνίο-
ῦ · Δάρνος προάγων ·
15 Τρόπιμος 'Ιταλικοῦ ·
[Κ]αλλικλῆ[ς] 'Ιταλικοῦ ·
['Ι]ταλικὸς Κ[α]σίου ·
[Γ]άϊος τρίς ·
['Ρ]ουπῖνο(ς) Σωκράτο[υς]
20 ['Α]λέξανδρος δίς ·
[Ζ]ώσιμος Διασκουρίδου ·
'Απολλώνιος Τυδραηον? ·
'Απολλώνιος γ' Παρμόν[ου] ·
[Δι]ασκουρίδης Δι[ασκουρίδου?]
25 ["Α]τταλος γ' 'Ερμέου ·
'Ηρακλίδης Νεάρκου ·

Nos. 44–45.

Karamanli.   Quadrangular basis originally surmounted by a round column now broken off. In the cemetery. Impressions.

A.

ΑΓΑΘΗΤΥΧΗΕΤΟΥΣ∤

ΟΙΜΥΣΤΑΙΤΟΥΔ
ΟΥΥΠΕΡΣΩΤΗΡ
ΤΟΥΔΗΜΟΥΟΡΛ
5 ΤΗΡΙΑΣΑΝΝΙ
ΚΑΙΤΙΒΕΡΙΟΥ
ΠΙΤΡΟΠΟΥ
ΠΡΑΓΜΑ–
ΤΟΥΚΑΙΛ
10 ΚΕΛΛΙΩ

B.

ΑΥΡΗΛΛΙΟΣ

ΚΙΔΡΟΛΛΑΣΤΡΙΣΙΕΡΕΥΣΔ
ΙΟΣΣΑΥΑΖΙΟΥΚΑΙΗΓΥ
ΝΗΑΥΤΟΥΑΡΤΕΜΕ
ΙΣ

5 ΥΡΗΛΛΙΟΣΑΥΡΗΛΛΙΟΣ
ΤΑΛΟΣΟΣΑΕΙΙΕΡΕ
ΛΛΟΥΟΣΑΕΙΣΑΤ

A.

'Αγαθῇ Τύχῃ · Ἔτους [τβ'?]
Οἱ μύσται τοῦ Δ[ιὸς Σαναζί-]
ου ὑπὲρ σωτηρ[ίας αὐτῶν καὶ]
τοῦ δήμου 'Ορ[μηλέων καὶ σω-]
5 τηρίας 'Αννί[ας Φαυστείνης]   ·
καὶ Τιβερίου [Κλαυδίου ἐπὶ ἐ-]
πιτρόπου [Κριτοβούλου, ἐπὶ]
πραγμα[τευτῶν 'Αβασκάν-]
του καὶ ['Ανθίνου καὶ Μαρ-]
10 [κ]ελλίω[νος].

B.

Αὐρήλλιος
Κιδρολλᾶς τρὶς ἱερεὺς Δ-
ιὸς Σαναζίου καὶ ἡ γυ-
νὴ αὐτοῦ Ἄρτεμεις.   ·
5 [Α]ὐρήλλιος ⟨Αὐρή[λ]λιος⟩
['Ατ]ταλος 'Οσαεὶ · Ἱερε-
[ὺς 'Αττά]λου · 'Οσαεὶς 'Ατ-
[τάλου].

This and the following inscription restore each other mutually. Concerning Zeus Sabazios see Foucart, *Les Associations religieuses chez les Grecs*, p. 77 ff. ; Ἐφημερὶς Ἀρχαιολογικὴ, 1883, p. 245 ff. publishes an inscription from the Piraeus concerning the worshippers of the God.

No. 46.

Karamanlü, In front of the Mesdjid. Bulletin de Correspondance Hellénique, 1878, *p.* 243. *Impression.*[1]

---

[1] Ligatures occur: line 4, NH; line 15, MH; line 17, MH; line 20, HN; line 21, MH; line 22, HN, MH; line 23, MHN; line 24, MHN; line 25, MHN; line 26, MHN; line 27, MHN; line 28, MHN. The close of line 23 seems to be MHNIΔ with MHN in ligature, but it is not absolutely certain. In line 10, the C in ABACKAN is small, having been originally omitted by the stonecutter.

```
▨▨▨▨▨▨IOΣΣAOYAZ
▨▨▨▨▨IAΣAYTΩNKAI
▨▨▨▨HΛEΩNKAIΣΩ
▨▨▨▨AΣΦAYΣTEINHΣ
```
5
```
▨▨▨KΛAYΔIOYEΠIE
▨▨▨KPITOBOYΛOYEΠI
▨▨▨EYTΩNABAΣKAN
▨▨▨NΘINOYKAIMAPⁱ
▨▨▨NOΣEΠIMIΣΘΩTΩN
```
10
```
▨▨▨IOYABAᵒKANTOYKAI
▨▨▨OΣNEIKAΔOYHPA
▨▨▨ΔOYKAINEIKAΔOYΔIΣ
▨▨▨TEYONTOΣKIΔPAMA
▨▨▨ΞΔIΣKAIEΛΠIΔOΣTHΣ
```
15
```
▨▨▨NAIKOΣAYTOYMHNIΣΔI
▨▨▨HΔOYOΣAIΣATTAΛOYIE
▨▨▨YΣEPMOYMAPKOΣΔIΣMH
▨▨▨OΓENOYENΘEOΣKPATEPO
▨▨▨YMAXOYIEPEYΣΔIONYΣOY
```

Var. Lect.

1. The *Bulletin* omits I in init. and Z in fine.
2. "   "   reads IAAΣ in init. and KA in fine.
3. "   "   omits Ω in fine.
6. "   "   omits I in fine.
8. "   "   reads NΘ in init. and AP▨ in fine.
9. "   "   indicates a missing letter in fine.
11. "   "   indicates a break in fine.
12. "   "   reads ΔIO in fine.
14. "   "   reads ΔI in init.
15. "   "   indicates a break in fine.
16. "   "   reads ⌐Δ in init.
18. "   "   indicates no break in init., and a break in fine.
19. "   "   indicates no break in init.

```
20  ▨HNIΣATTAΛΟΥΜΟΥΝΔΙΩ
    ▨ΟΣΑΤΤΑΛΟΣΔΗΜΗΤΡΙΟΥ
    ▨HNIΣΑΤΕΙΜΗΤΟΥΙΣΠΑΤΑ
    ▨ΟΣΜΗΝΙΔΟΣΕΝΘΕΟΣΜΗΝΙΔ
    ▨ΣΜΗΝΙΣΠΟΣΙΔΩΝΙΟΥΕΝΘ
25  ▨ΣΑΚΕΠΤΟΣΜΗΝΙΔΟΣΟΣΑΙ▨
    ▨ΛΑΥΚΟΥΜΗΝΙΣΝΕΙΚΟΛΑΟΥ
    ΚΙΒΥΡΑΤΟΥΜΕΝΙΣΤΕΥΣΜΗΝΙ
    ▨ΟΣΜΕΝΙΣΤΕΟΣΝΕΑΡΧΟΣΜΗΝΙΔ
    Ο▨ΝΕΑΡΧΟΣΝΕΙΚΟΛΑΟΥΚΙΒΥΡΑΤ▨
```

Var. Lect.

20. The *Bulletin* reads ꓱN in init., indicating no break for the missing M.
21. The *Bulletin* indicates no break in init.
22. " "   reads ꓱN, indicating no further break in init.
23. " "   indicates no break in init., closing with ΕΝΘΕΟΣ.
24. " "   omits entirely.
25. " "   indicates no break in init., and reads ΑΕ in fine.
26. " "   indicates no break in init.
28. " "   reads ΟΣ, and does not indicate a break in init.; but does indicate a break in fine.
29. The *Bulletin* reads ΟΝ in init., indicating no break between O and N; it reads ΤΟ▨ in fine.

['Αγαθῇ Τύχῃ · ᾿Ετους τβ?]
[Οἱ μύσται τοῦ Δ]ιὸς Σαουαζ[ίου]
[ὑπὲρ σωτηρ]ίας αὐτῶν καὶ
[τοῦ δήμου ᾿Ορμ]ηλέων καὶ σω[τ-]
[ηρίας ᾿Αννί]ας Φαυστείνης
5 [καὶ Τιβερίου] Κλαυδίου ἐπὶ ἐ-
[πιτρόπου] Κριτοβούλου, ἐπὶ
[πραγματ]ευτῶν ᾿Αβασκάν-

[του καὶ 'Α]νθίνου καὶ Μάρ[κ-]
[ελλίω]νος, ἐπὶ μισθωτῶν
10 [Κλαυδ]ίου 'Αβασκάντου καὶ
[Μήνιδ]ος Νεικάδου 'Ηρα-
[κλεί]δου καὶ Νεικάδου δὶς,
[ἱερα]τεύοντος Κιδράμα-
[ντος] δὶς καὶ 'Ελπίδος τῆς
15 [γυ]ναικὸς αὐτοῦ· Μῆνις Δι-
[ομ]ήδου, 'Οσαὶς 'Αττάλου ἱε-
[ρε]ὺς 'Ερμοῦ, Μάρκος δὶς Μη-
[ν]ογένου, "Ενθεος Κρατερο-
[ῦ Σ]υ(μ)μάχου ἱερεὺς Διονύσου,
20 [Μ]ῆνις 'Αττάλου Μουνδίω-
[ν]ος, "Ατταλος Δημητρίου,
[Μῆ]νις 'Ατειμήτου (δ)ὶς, Πάτα-
[λ]ος Μήνιδος, "Ενθεος [Μήνιδ-]
[ος], Μῆνις Ποσιδωνίου, "Ενθ-
25 [εο]ς "Ακεπτος Μήνιδος 'Οσαὶ
[Γ]λαύκου, Μῆνις Νεικολάου
Κιβυράτου, Μενιστεὺς Μήνι-
[δ]ος Μενιστέος, Νέαρχος Μήνιδ-
ο[ς], Νέαρχος Νεικολάου Κιβυράτ[ου].

Nos. 47–50.

Karamanlü. Broken quadrangular cippus in the cemetery. The top moulding is broken entirely away. The whole present height of the stone is 1.09 m.; to the bottom moulding, 0.90 m.; width, 0.61 m. Bulletin de Correspondance Hellénique, 1878, p. 257 ff. *Impressions.*[1]

[1] Ligatures occur: line 13, HN; line 16, HN; line 17, MH; line 19, ШN, MH; line 22, MH; line 24, MH. The N at the end of line 27 is the numeral belonging to line 28, but is written for reasons known only to the stonecutter above the ✳.

A.

ᴄ̇ʟ̇ʏ̇ᴏ N   O X Λ C⧰

✱ N -⊳ ·   [uncut]

A Κ Є Π Τ Ο C   Μ Η Ν Ι

5   Δ Ο C Є Τ Є Ι   Μ Η C Є Ν

Τ Ο Ν   Ο Χ Λ Ο Ν

✱ N   [uncut]

Α Γ Α   Θ Ο Π Ο

Υ C Α Τ   Τ Η Є C

10   Τ Є Π   Α Ν Ш

C Є ⩰ Κ Є

[Horseman]

Λ Τ Τ Α Λ Ο Ċ Κ Α Ι Α Μ Υ Ν Τ Α C

Μ Η Ν Ι Δ Ο C Κ Α Λ Α Μ Ι C Κ Ο Υ

Є Τ Ι Μ Η C Α Ν Τ Ο Ν Ο Χ Λ Ο Ν

15   ✱ N

⧰Μ Η Ν Ι C Ν Є Ι Κ Α Δ Ο Υ Π Ο Λ Υ

Δ Є Υ Κ Ο Υ Μ Є Τ Α Δ Η Μ Η Τ Ρ⧰

Ο Υ Τ Ο Υ Α Δ Є Λ Φ Ο Υ Κ Α Ι Ш

Ν Υ Ι Ш Ν Є Τ Ι Μ Η C Є Ν Τ Ο Ν Ο

20   ⧰Λ Ο Ν ✱ C Ο Є

⧰C Α Є Ι C Α Τ Τ Α Λ Ο Υ ✱ N

Var. Lect.

1. The *Bulletin* reads ∟ ı ᴜ ɴ in init., and omits OXΛC⧰ in fine.
4.   "   "   reads MHVı in fine.
5.   "   "   reads MICЄN in fine.
9.   "   "   reads YCA   THЄY.
11.   "   "   reads CI✱.
12.   "   "   reads A in init.
16.   "   "   indicates no break in init.
20.   "   "   indicates no break in init.
21.   "   "   does not note that the ✱N in fine belong to the
end of 22, being engraved above the line, as often happens.

```
     ЄΤЄΙΜΗCЄΝΤΟΝΟΧΛΟΝ
     ΟΝΗCΙΜΟCΜΗΝΙΔΟCЄΤЄΙ
     ΜΗCЄΝΤΟΝΟΧΛΟΝ✳ΚЄ
  25 ΜΗΝΙCΤΡΙCΜΙΛΛΑΚΟC
     ЄΤΙΜΗCЄΤΟΝΟΚΛΟΝ
     ΜΑΡΚΟCΔΙCЄΝΘЄΟC   Ν
     ЄΤΙΜΗCЄΝΤΟΝΟΧΛΟΝ✳
     ΛЄΥΚΙCΜΗΝΙCЄΤЄΙΜΗCЄΤ
  30 ΟΧΛΟΝ✳ΚЄ
     ΜΗΝΙCΔΡΑΥΚШΝЄ
     ΗCЄΝΤΟΝΟΧΛΟΝ
     ΑΓΑΘΟΠΟΥCΜΗ
     ΛΙΛCΚΑΝΤΟΥЄΤ
  35 ΝΤΟΤΤΟΚΛΟΝ
```

Var. Lect.

23. The *Bulletin* reads · · ΙΜΟC in init.
24. " " reads ✳M(?) in fine.
25. " " reads ΜΗΝΙCΦΙ and nothing more.
26. " " reads · ΤΙΜΗCЄΝ · · Ν.
27. " " reads ЄΝΟC in fine.
28. " " reads · ΤΙ in init. and ΧΛΟΝ in fine, failing to give the numeral Ν above the line.
29. The *Bulletin* reads ΛΟΥΚΙC   ЄΤЄΙΜΗCЄΝ.
31. " " reads ΜΗΝΙCΔΙ · ΜΟΝЄ.
34. " " reads · · ΛCΚ.
35. " " reads ΝΤΟΝΟΧΛΟΝ.

B.

A fragment, — lines 14 to 19, — which has been broken from the stone, fits in as indicated in the text.[1]

---

[1] Lines 14 to 19 represent the fragment given in the *Bulletin de Correspondance Hellénique*, 1878, p. 259. Ligatures occur: line 7, HN; line 9, HN.

```
        ▨▨▨ΛΑΙΟΥΖΕϹΖ
        ΝΟΤΟΚΟϹΕΙϹΛΑΧ▨
        ΝΟΝΧШΡΗϹΕΙ
        ΜΗΝΙϹΤΡΙϹΜΕΛΙϹ
   5    ϹΟΡΓΟΥΕΤΕΙΜΗϹΕ
        ΤΟΝΟΧΛΟΝ✶Ν
        ΜΗΝΙϹΛΤΤΑΛΟΥΚΙΚΚΟΥΕΤ▨
        ΜΗϹΕΤΟΝΟΧΛΟΝΕΛΑΙΟΥΖϹ
        ϹΤΑϹΖΜΗΝΙϹΒΙΛΛΙΟΥϹ▨
 · 10   ΤΕΙΜΗϹΕΤΟΝΟΧΛΟ▨
        ϹΟΥΡΝΟϹΝΕΟϹΕ̅▨
        ΤΟΧΟΧΛΟΝ✶Ν▨
        ▨ΗΤΡΙΟΥϹ̅̅▨
        ▨▨▨▨ Ε ▨▨
  15    ▨▨▨▨ϹΙΟΥ▨
        ▨▨ ✶ΚΕ ▨▨
        ▨▨ΕΤΙΜΗ▨▨
        ▨▨ΠΕΡΤΟΥ▨
        ▨▨ΟΥΡ▨
  20    ▨▨▨▨▨▨
```

Var. Lect.

2. The *Bulletin* reads ΙΟΤ in init., and ΧΑ in fine.
8.   "   ·   "   omits Ϲ in fine.
9.   "   "   reads ΜΗΝΙϹΡΙΜΙϹ in fine.
11.   "   "   reads ΝΕΑ▨ in fine.
13.   "   "   reads ▨▨▨ ΟΥ▨.

Lines 63–67 of the *Bulletin* belong to this inscription, as is perfectly clear when the fragment is adjusted to the stone. Hence

14. The *Bulletin* reads ▨Ε.
15.   "   "   ▨ΥϹΙΟ▨.
16.   "   "   omits entirely.

C.[1]

```
▨▨▨▨▨▨▨▨          Δ Є Π Α
⊒ A C ✳ K Є        K A I O I
N O Y Ƶ Є C        T A C P̄
M H N I C          K A C Ṫ O
5  P O C Є T        Є I M H
Ƨ E N T             O N O
X Λ O N            ✳ K Є
M H N I Ϲ A Π ◻·Λ    Λ Ш N I ◻ Y Δ A
P Ш N ◻ Ϲ E T E I M H Ϲ E ✳ Λ Z
10 M Є N I Ϲ T Є Y C M H N I Δ O C Λ Y Γ O C
T P O Π O C Є T Є I M H C Є N T O N O X Λ O N'
▨▨▨▨▨▨▨▨▨▨▨▨▨▨ T ▨▨
▨▨M H C Є N T O N O K Λ O N▨▨
▨ H M H T▨Y Ϲ T P A T▨▨
15 ▨ Ϲ N ◻ N H X Λ ▨▨▨
▨▨ Π ▨ Λ Ш N I O C ▨▨
▨▨ T I M I ▨▨
```

[Here follow nine hopelessly defaced lines.]

Var. Lect.

2. The Bulletin reads AC in init.
3.   "   "   reads TAC · P in fine.
12.   "   "   reads · · · KЄ.
13–17. The Bulletin omits entirely, with the remark: "Cette face n'est pas remplie."

---

[1] In line 4 the T is inserted above the line between C and O. The N at the end of line 11 is incised immediately above the O. In line 13 the letters TONOKΛON are small and somewhat cramped. Ligatures occur: line 8, HN; line 13, HMH; line 15, ◻NH, the reading of this line is very doubtful.

$D.$[1]

KΑCIOCKAIKPATE
YIOIΓAIOYMHNIΔ
ETEIMHCANTON
YPIBIKAITON
5   OIAYTOIKACIO
TEPOCETIE
NΩCAN✳N
CIOCΔICK
MHCEN
10

Var. Lect.

1. The *Bulletin* reads · · · · · · · · CKAI in init.
2.   "   "   reads YIOI · AI in init., and IΖ in fine.
3.   "   "   reads ·T in init.
4.   "   "   reads TOPIBI · KAITO.

6–9 are entirely omitted by the *Bulletin*.

A.

In lines 3, 7, 11, 15, etc., I read δηνάρια in the Accusative, because the Accusative ξέστας is certain in inscription *B*, line 9, and in inscription *C*, line 3.

['Ο δεῖνα τοῦ δεῖνος ἐτίμησε τ]ὸν ὄχλ[ον]
(δηνάρια πεντήκοντα)·
῎Ακεπτος Μήνι-
5   δος ἐτείμησεν
τὸν ὄχλον
(δηνάρια πεντήκοντα)·

---

[1] Ligatures occur: line 2, MHN; line 3, MH.

Ἀγαθόπο-
υς Ἄττη ἐσ-
10 τεπάνω-
σε (δηνάρια εἴκοσι καὶ πέντε)·
[Ἄ]τταλος καὶ Ἀμύντας
Μήνιδος Καλαμίσκου
ἐτίμησαν τὸν ὄχλον
15 (δηνάρια πεντήκοντα)·
[Μ]ῆνις Νεικάδου Πολυ-
δεύκου μετὰ Δημητρ[ί]-
ου τοῦ ἀδελφοῦ καὶ τῶ-
ν υἱῶν ἐτίμησεν τὸν ὄ-
20 [χ]λον (δηνάρια διακόσια ἑβδομήκοντα πέντε)·
['Ο]σαεὶς Ἀττάλου
ἐτείμησεν τὸν ὄχλον (δηνάρια πεντήκοντα).
Ὀνήσιμος Μήνιδος ἐτεί-
μησεν τὸν ὄχλον (δηνάρια εἴκοσι καὶ πέντε)·
25 Μῆνις τρὶς [Μ]ίλλακος
ἐτίμησε τὸν ὄκλον. [(δηνάρια . . . . ?)]
Μάρκος δὶς Ἔνθεος
ἐτίμησεν τὸν ὄχλον (δηνάρια πεντήκοντα)·
[Λ]εύκις Μῆνις ἐτείμησε τ[ὸν]
30 ὄχλον (δηνάρια εἴκοσι καὶ πέντε)·
[Μ]ῆνις Δραύκων ἐ[τείμ-]
ησεν τὸν ὄχλον [(δηνάρια) . . . . . ?]
Ἀγαθόπους Μή[νιδος]
['Αβ]ασκάντου ἐτ[είμησε-]
35 ν τὸ[ν ὄκλον [(δηνάρια) . . . . . ?]

B.

[Ὁ δεῖνα ἐτείμησεν τὸν ὄχλον
ἐ]λαίου ξέσ(τας ἑπτὰ) [ᾧ-

ν] ὁ τόκος εἰς λάχ[α-]
νον χωρήσει ·
Μῆνις τρὶς Μελισ-
5 σόργου ἐτείμησε
τὸν ὄχλον (δηνάρια πεντήκοντα) ·
Μῆνις ['Α]ττάλου Κίκκου ἐτε[ί-]
μησε τὸν ὄχλον ἐλαίου ξ[έ-]
στας (ἑπτὰ) · Μῆνις Βιλλίου [ἐ-]
10 τείμησε τὸν ὄχλο[υ (δηνάρια) . . . . . ?]
Σοῦρνος νέος ἐ[τείμησε]
τὸ(ν) ὄχλον (δηνάρια πεντήκοντα) · [ὁ δεῖνα]
[Δημ]ητρίου [ἐτείμησεν τὸν ὄχλον κ.τ.λ.]
['Ο δεῖνα ἐτ]ε[ίμησεν κ.τ.λ.]
15 ['Ο δεῖνα Διον]υσίο[υ ἐτείμησεν]
[τὸν ὄχλον] (δηνάρια εἴκοσι καὶ πέντε) ·
['Ο δεῖνα ἐτίμη[σεν κ.τ.λ.]
. . . . . ὑ[πὲρ τοῦ . . . . .
. . . . ουρ . . . . .

C.

. . . .
. . . . . (δηνάρια εἴκοσι καὶ πέντε) καὶ οἴ-
νου ξέστας (ἑκατόν) ·
Μῆνις Κάστο-
5 ρος ἐτείμη-
σεν τὸν ὄ-
χλον (δηνάρια εἴκοσι καὶ πέντε) ·
Μῆνις 'Απολλωνίου Δά-
ρωνος · ἐτείμησε (δηνάρια τριάκοντα ἑπτά) ·
10 Μενιστεὺς Μήνιδος λυγοσ-
τρόπος ἐτείμησεν τὸν ὄχλον [κ.τ.λ.]
['Ο δεῖνα τοῦ δεῖνος ἐ]τ-

[εἰ]μησεν τὸν ὄκλον [κ.τ.λ.]

[Δ]ημήτ[ριος] Στράτ[ωνος ἐτείμ-]

15 [ησ]εν [τ]ὸν [ὄ]χλο[ν κ.τ.λ.]

['Α]π[ολ]λώνιος [τοῦ δεῖνος]

[ἐ]τίμ[ησεν τὸν ὄχλον κ.τ.λ.]

Line 10. λυγοστρόπος is probably an appellative = λυγοστρόφος, a withe-twister.

D.

Κ[ά]σ[ι]ος καὶ Κρατε[ρὸς]

υἱοὶ Γαΐου Μήνιδ[ος]

ἐτείμησαν τὸν [ὄχλον]

[✻] ριβ ⟨I⟩ καὶ τὸ[ν] τόκον.

5 Οἱ αὐτοὶ Κάσιο[s καὶ Κρα-]

τερὸς ἔτι ἐ[στεφά-]

νωσαν (δηνάρια πεντήκοντα)· [Κά-]

[σ]ιος δὶς Κ[ρατεροῦ ἐτεί-]

μησεν [τὸν ὄχλον κ.τ.λ.]

No. 51.

Karamanlii. Fragment in the court of a house.

No. 52.

Karamanlii. Cippus serving as a basis to a pillar of the vestibule of the Mcsdjid. Length, 1.02 m.; length within the mouldings, 0.71 m.; width, 0.32 m. Copy and impression.[1]

---

[1] Lines 1–4 are on the top moulding. ·Ligatures occur in lines : 2, NH̆; 4, MH.

```
                ..  Ο Ϲ Ο Γ Μ Η Λ
     Κ Λ Η Ρ Ο Ν Ο Μ Ω Ν Φ Α Υ Ϲ Τ Ε Ι Ν Η
              Ϲ Ο Υ Φ Α Υ Ϲ Τ Ε Ι Ν Η Ϲ Ο Υ

     Μ Η Λ Ι Α Ϲ Κ Ο Ρ Ν Ο Φ Ι Κ Ι Α Ϲ

  5  Α Υ Ρ Α Ρ Τ Ε Ι Μ Η Ϲ Χ Α Ρ Η
     Τ Ο Ϲ Μ Ο Γ Γ Ο Υ Ε Τ
     Ε Ι Μ Η Ϲ Ε Ν Τ Ο Ν Δ Η
     Μ Ο Ν Α Ρ Ι Ϲ Τ Ο Ν Κ Ι
     Ι Α Τ Τ Ι Κ Α Ϲ Τ Ο
 10  Μ Η Ν Ι Ϲ Κ Ε Α Ρ Τ Ε Ι Μ Η Ϲ
     Ο Ι Υ Ι Ο Ι Α Ρ Τ Ι Μ Ο Υ Χ Α Ρ Ν
     Δ Ο Ϲ Μ Ο Υ Ν Γ Ο Υ Ε Ϲ Τ Ε
     Π Α Ν Ω Ϲ Α Ν Τ Ο Ν Δ Η Μ
     Ο Ν ✳ Ϲ Ω Ν Ο Τ Ο Κ Ο Ϲ Υ
 15  Κ Ω Ρ Η Ϲ Ι Κ Α Τ Ε Τ Ο
     Ϊ Ε Ι Ν Ο Μ Ε Ν Ο Τ Ο Υ
     Κ Υ Τ Ο Υ
```

..... [δῆμ]ος Ὀ[ρ]μη[λέων]
κληρονόμων Φαυστείν[η-]
ς [θ]υ(γατρὸς) Φαυστείνης Οὐ[μ-]
μη[δ]ίας Κορνοφικίας·
5 Αὐρ. Ἀρτειμῆς Χάρη-
τος Μόγγου ἐτ-
είμησεν τὸν δ[ῆ-]
μον ἄριστον κ[α-]
ὶ ἀ[νε]ικαστ[ότατον] ....
10 [Μ]ῆνις κὲ Ἀρτειμῆς
οἱ υἱοὶ Ἀρτίμου Χάρ[μι-]
δος Μούνγου ἐστε-

πάνωσαν τὸν δῆμ-

ον (δηνάρια διακόσια), ὧν ὁ τόκος [ὑπ-]
16 [ο]κωρῆσι κατ᾿ ἔτ[ος ὁ]
[γ]εινόμενο[ς] τοῦ [ἀρ-]
[γ]ύ[ρ]ου.

Line 3. I lay no stress on the conjecture of θυγατρὸς; but if it be possible, it helps out of a difficulty. Οὐμμηδίας is miswritten for Οὐμμιδίας. The meaning of lines 14–17 is that the money shall be funded and the interest expended yearly.

June 8. Karamanlü to Tefeny, 1 h. 3 m. At Tefeny I again met Messrs. Ramsay and Smith. Here we copied the following inscriptions.

Nos. 53-55.

Tefeny. Quadrangular cippus in the cemetery. Height, 1.58 m.; within mouldings, 1.20 m.; width, 0.41 m. Bulletin de Correspondance Hellénique, 1878, *p.* 56 ff.; C.I.G. 4366 w. *The inscriptions are so badly blurred by the gradual wearing away of the stone that impressions would be worthless. The reader will have to accept the texts on the united testimony of Mr. Ramsay and myself.*

A.

Lines 1–25 copied by J. R. S. S., copy verified by W. M. Ramsay; lines 26–36 copied by W. M. Ramsay, copy verified by J. R. S. S.[1]

---

[1] Ligatures occur: line 5, the second NE; line 6, NE; line 7, NE, HP; line 10, HP; line 14, HP, NE; line 18, MHN; line 19, MH; line 20, NE, HP; line 21, MHN; line 22, MHN; line 23, MHN; line 24, NE, MHN, NE; line 25, NE; line 28, NE; line 29, NE; line 30, MHN; line 31, MHN, MHN, NE; line 32, NE, NE; line 33, NE. In line 9, OYP occurs twice, and in both cases the Y is written above the line between O and P. In line 13 a small N is inserted between Y and Δ. In line 21 the Y of OYΛ is inserted above the line between O and Λ. In line 24 the OY at the end of the line is written above AΔ. In line 32 the O at the end of the line is written above the Δ.

ΤΗΡΙΑΣΑΝΝΙΑΣΦΑΥ▨
ΤΕΙΝΗΣΚΑΙΔΗΜΟΥΟΙ▨
ΜΗΛΕ▨ΝΕΠΙΑΒΑΣΚΑ▨
ΤΟΥΠΡΑΓΜΑΤΕΥΤΟΥ
5 ΝΕΙΚΟΛΑΟΣΝΕΑΡΧΟΥ
ΝΕΙΚΟΛΑΟΣΣΥΜΜΑΧΟΥ
ΝΕΙΚΑΔΑΣΗΡΑΚΛΕΙΔΟΥ
ΑΠΟΛΛΟΔΟΤΟΣΜΙΛΛΑΚΟΣ
ΚΑΛΠΟΥΡΝΙΣΣΟΥΡΝΟC
10 ΜΗΝΙΣΝΕΙΚΑΔΟΥΗΡ▨
ΜΗΝΙΣΟΣΑΕΙ   ΚΛΕΙΔΟΥ
ΜΗΝΙΣΑΠΟΛΛΩΝΙΟΥ
ΑΤΤΑΛΟΣΜΟΥΝΔΙΩΝΟΣ
ΗΡΑΚΛΕΙΔΗΣΝΕΙΚΑΔΟΥ
15 ΚΡΑΤΕΡΟΣΣΥΜΜΑΧΟΥ
ΟΣΑΕΙΣΑΤΤΑΛΟΥ

Var. Lect.

1. The *Bulletin* reads ΙΑΣΑΝ, and adds a Σ in fine, indicating no break.
2. The *Bulletin* reads ΟΡ in fine.
3. "   "   reads ΜΗΛΕΩΝ in init.
5. "   "   reads ▨ΜΟΣΝΕΑΡΧΟΥ.
6. "   "   reads ▨ΟΣΣΥΜΜΑΧΟΥ.
7. "   "   reads   ΛΔΑΣΗΡΑΚΛΕΙΔΟΥ.
8. "   "   reads · · Ο · · ΟΔΟ · ΟΣ · · · ΛΛΑΚΟΣ.
9. "   "   reads ΚΑΛΠΟΡΝΙΣΣΟΡΝΟΣ.
10. "   "   reads ΗΡ▨ in fine, failing to note the ΚΛΕΙΔΟΥ immediately below the line.
11. The *Bulletin* reads ▨ΝΙΣΟΣΑΕΙ▨.
12. "   "   reads ▨ΟΛΛΙΑΝΙΟΥ.
13. "   "   reads ΑΤΤΑΛΟΣΛ▨ΔΙΩΝΟΣ.
14. "   "   reads ΗΡΑ · · · ΔΗ · ΝΕΙΚΑΔΟΥ·
16. "   "   reads ΟΡΣΑ in init.

ΜΑΡΜΑΣΜΗΝΙΔΟΣ
ΧΑΡΗΣΜΗΝΙΔΟCΜΑΡΜΟΥ
ΜΗΝΙΣΤΡΙΣΜΕΛΙΤΩΝ
20   ΝΕΙΚΑΔΑΣΔΙΣΗΡΑΚΛΕΙΔΟΥ
ΚΑΡΠΩΝΜΗΝΙΔΟΣΟΥΛΟ░
ΚΡΑΤΕΡΟΣΜΗΝΙΔΟΣΔΙΔΥΜΟΥ
ΚΑΛΛΩΝΜΗΝΙΔΟΣΜΑΡΜΟΥ
ΝΕΙΚΑΔΑΣΜΗΝΙΔΟΣΝΕΙΚΑΔΟΥ
25   ΝΕΙΚΑΔΑΣΗΡΑΚΛΕΙΔΟΥ
ΜΗΝΙΣΚΑΛΑΜΙΣΚΟΥ
ΜΗΝΙΣΔΙΟΜΗΔΟΥ
ΔΙΟΝΥΣΙΟΣΝΕΙΚΟΛΑ░
░ΥΜΜΑΧΟΣΝΕΙΚΟΛΑ░
30   ░ΝΙΣΜΗΝΙΔΟΣΟΣΑΕΙ
ΜΗΝΙΣΜΗΝΙΔΟΣΝΕΙΚΑΔΟΥ
ΝΕΙΚΑΔΑΣΝΕΙΚΑΔΟΥΗΡΑΚΛΕΙΔΟ░
░ΠΟΛΛΟΔΟΤΟΣΝΕΙΚΟΛΑΟΥ
ΜΟΛΥΖΚΑΣΤΟΡΟC
35   ░ΟCΜΕΝ░ΝΔΡΟΥ
░CΧΑΡΗΔΟCΜΟ
░░░░░░░

Var. Lect.

17. The *Bulletin* reads ΜΑΡΜΑΣ ·· ΝΙΔΟΣ.
19.   "   "   indicates a break in fine.
20.   "   "   reads · ΕΙΚΑΔΑΣΣΙΣ in init.
21.   "   "   reads ΧΑΡΙΩΝΜΗΝΙΔΟΣΟΛΟ░.
23.   "   "   reads · ΩΛΙΩΝ in init.
24.   "   "   reads ░░ΔΑΣΜΗΝΙΔΟΣΝΕΙΚΙ░.
28.   "   "   reads ΛΑΟ · in fine.
30.   "   "   omits entirely.
31.   "   "   reads ·· ΝΙΣ in init.
32.   "   "   reads ΕΙΔ░ in fine.
35.   "   "   reads ░ΘΟΝΟΝΗ · ΝΑΤΟΥ.
36.   "   "   reads ΜΕ░ in fine.

B.

Copied by J. R. S. S.; copy verified by W. M. Ramsay.[1]

```
     ΜΗΝΙΣΜΗΝΙΔΟΣΟΥΑΛ
     ΚΛΑΥΔΙΟΣΚΡΑΤΕΡΟΥ
     ΣΥΜΜΑΧΟΣΚΡΑΤΕΡΟΥ
     ΑΡΙΣ░░ΠΟΣΝΕΙΚΟΛΑΟΥ
  5  ΔΙΟΝ░ΣΙΟΣΜΗΝΙΔΟΣ
     ΚΡΑΤΕΡΟΣΕΛΙΟΥ
     ΚΑΣΤΩΡΜΗΝΙΔΟΣ
     ΔΙΟΜΗΔΗΣΜΗΝΙΔΟΣ
     ΜΗΝΙΣΧΑΡΗΤΟΣΗΡΑΚΛΕΙΔΟΥ
 10  ΣΟΥΡΝΟΣΔΙΣΝΕΟΣ
     ΑΤΤΑΛΟΣΚΑΛΛΙΚΛΗΔΟΣ
     ΑΤΤΑΛΟΣΔΗΜΗΤΡΙΟΥ
     ░ΒΑΣΚΚΑΝΤΟΣΑΒΑΣΚΑΝΤΟΥ
```

Var. Lect.

1. The *Bulletin* reads ΟΥΛ░ in fine.
3. " " reads ΣΥΜ · ΑΧΟΣΚΡΑ░.
4. " " reads ΚΡ░ΣΝ · · ΚΟΛΛ░.
5. " " reads ΔΙΟΝΥ░ΟΣΜ░.
6. " " reads Κ · · · ΔΙΟΣ░ΙΟ░.
7. " " reads ΧΑ░ΜΗΝΙΔΟΣ.
9. " " reads ΑΝΗΣΧΑΡΗ · ΟΣΗ░.
10. " " reads ΝΕΟΖ░ in fine.
11. " " reads ΛΗΟϹ in fine.
13. " " reads ΤΑΜΑΝΤΟΣ░.

---

[1] Ligatures occur: line 1, HN, HN; line 4, NE; line 5, MH; line 8, HϹ, MHN; line 9, MHN, HP; line 14, MHN; line 17, HN; line 19, HN; line 20,˙MHN; line 21, ░MHN, MH; line 24, NE, MH; line 25, HN, NE, MH; line 28, NE; line 29, MH; line 32, HϹ; line 34, ░NE. In line 15 the NEOϹ at the end of the line is written in smaller letters than those in the rest of the line. In line 19 the letters P░OY are written above the line. Between lines 28 and 29 there is a blank line which was never incised.

MHNICKAΛΛIKAHΔOΣ
15 ΛIKINNIOΣΔIΣΝεοϲ
▨ΕNANΔPOΣKPATEPOY
MHNIΣΣYMMAXOY
ATTAΛOΣOΣAEI
MHNIΣATTAΛOYΔHMHTP▨OY
20 ΚAΛIKAHCKAΛΛIKΛEOYCMHNIΔOC
▨MHNIΣMHNIΔOΣΔIOMHΔOY
MHNIΣΔICTOYMIΛAKOΣ
ΓNAIOΣMHNIΔOΣ
KΛ·NEAΓXOCMHNIANOY
25 MHNICNEIKAΔOYMHNIΔOC
MHNICATTHΔOC
ATT/·ΛOCΔICΔHMHTPIOY
▨ΕNANΔPOΣNEAPXOY
▨ΤΣTEACMHNIΔOCMIΛΛAKOΕ

Var. Lect. .

14. The *Bulletin* reads ▨▨▨KAΛΛIKAH▨.
15. "   "   reads ▨▨HNIΔOΣΔIΣ.
17. "   "   reads · HN in init.
18. '"   "   reads ▨▨∠ OYˉAΣ.
19. "   "   reads · HNIΣΛTTAΛOYΔ▨.
20. "   "   reads KΛEOYˉMHNIΣ in fine.
21. "   "   reads · HN in init. and HΔOY in fine.
22. "   "   reads MHNIΣΔIΣ▨MIΛANOΣ.
24. "   "   reads ▨N▨ΔIXOLMHNIΔIO.
25. "   "   reads ▨▨KAΔOYMHIΔOΣ.
26. "   "   reads ▨▨▨▨▨NIΔOC.
27. "   "   reads OCAICΔHMHTPIOY.

After line 28 the *Bulletin* inserts a line as wholly defaced; no such line exists.

29. The *Bulletin* reads ▨MHNIΔOCΝΛΛΛKOC.

30 ▨ΝΤΩΝΙΟΣΔΙΣΚΥΝΑΚΟΣ
K·ΣΟΥΡΝΟΣΤΡΙΣ
ΚΑΡΗΣΓΝΑΙΟΥΤ▨ΚΤΩΝ
ΣΥΜΑΧΟΣ   ΜΑΝΟΥ
ΝΕΙΚΟΛΑΟΣΣΥΜΜΑΧΟ▨

35 ▨ΙΚΑΔΑΣΣΟΛΩΝΟΣΓΡΥΠΟΣ
ΛΕΝΑΝΔΡΟΣΣΟΥΡΙ▨
▨ΟΣΜΕΝΑΝΔΡΟ▨

Var. Lect.

30.	The *Bulletin* reads	▨ΕΚΥ▨.	
31.	" "	reads	▨ΡΝ▨.
32.	" "	reads ΚΑΙΗΣΓΝΑΙΟΥΤ▨ΩΝ.	
33.	" "	reads ▨ΜΜΑΧΟΣ▨ΜΑΚΟΥ (sic).	
34.	" "	reads ΝΕΙΚ in init., and ΣΥΜΑΧΟ▨ in fine.	
35.	" "	reads ΠΑΔΑΣΖΟΛΙΝΟΣΓΡΥΠΟΣ.	
36.	" "	reads ▨ΕΝ in init., and ΣΟΥΡ▨ in fine.	
37.	" "	reads ΔΡ▨ in fine.	

C.

Copied by W. M. Ramsay; copy verified by J. R. S. S.[1]

ΜΗΝΙΣΤΡΙΣΜΗΝΙΔΟΣ
ΝΙΚΑΔΟΥ [complete]
ΔΗΜΗΤΡΙΟΣΑΤΤΑΛΟ▨

Var. Lect.

1. The *Bulletin* reads ΜΗΝΙΣΜΗΝΙΔΟΣ.
2. " " reads ·· ΝΙΚΑΔΟΣ▨.
3. " " reads ΛΟΥ in fine.

---

[1] In line 15, ΜΗ are in ligature. In line 18 the ΟΣ does not belong to line 18, but serves as the final ending of ΜΕΛΙΤΩΝ in lines 17 and 19. The ΚΡΟΥ in line 20 belongs to the ΜΑ at the end of line 21. Line 32: Mr. Ramsay has a note that J. R. S. S. preferred to read ΚΑΙ at the beginning of the line.

```
   ΝΙΚΑΔΑΓΜΗΝΙΔΟΓΝΙΚΑΔΟΥ
·5 ΑΤΤΗΓΔΙΓΤΟΥΟΓΑΕΙ
   ΚΡΑΤΕΡΟΓΚΛΑΥΔΙΟΥ
   ΝΕΙΚΑΔΑΓΔΙΓΜ░ΝΙΔΟΓ
   ΚΑΜΕΝΑΝΔΡΟΓΓΟΥΠΝΟ░
   ΜΗΝΙΓΔΙΓΚΥΒΥΡΟΥ
10 ΜΗΝΙΓΜΑΚΡΟΥΜΙΛΛΑΚΟ░
   ΜΗΝΙΓΤΡΙΓΟΥΑΔΑΡΟΥ
 · ΚΑΛΠΟΥΡΔΑΟΓΓΟΥΠΝΟ
   ΟΓΑΕΙΓΜΗΝΙΔΟΓΟΥΑΔΑΡΟΥ
   ΚΛΑΥΔΙΓΔΙΓΚΡΑΤΕΡΟΥ
15 ΔΗΜΗΤΡΙΟΓΜΗΝΙΔΟΓΔΙΓΝΕΙΚΑΔΟΥ
   ΚΙΔΡΑΜΑΣΤΡΙΣ
   ░ΑΛΛΙΚΑΗΓΜΗΝΙΔΟΓΜΕΛΙΤΩΝ
   ΚΑΛ·ΓΟΥΠΝΟΓΔΑΟΥ      ΟΓ
   ΜΗΝΙΓΚΑΛΛΙΚΛΕΟΥΓΜΕΛΙΤΩΝ
20 ΜΗΝΙΓΤΡΙΓΚΙΒΥΡΟΥ      ΚΡΟΥ
   ΑΡΙΓΤΕΑΓΜΗΝΙΔΟΓΤΡΙΓΜΑ
```

Var. Lect.

5. The *Bulletin* omits as wholly defaced.
7. " " reads ΝΕΙΚΑΔΑΓΜ░ΙΔΟΥ.
8. " " reads ΚΑΜΕΡΙΑΜΑΡΟΓΓᶜΑΡΔΙΟΥ.
9. " " reads ΜΗΝΙΓΔΙΟΝΥΓΙΟΥ.
10. " " reads ΜΗΝΙΓΜΑΡΚΟ░ΕΛΛΑΚ░.
12. " " indicates a break in fine.
15. " " ΔΟΓΔ░ΙΚΑΔΟΥ in fine.
16. " " ΚΙΔΡΑΜΑΓΤΡΙΓ   ΚΑΛΛΙΚ░.
17. " " reads ΚΑΛΛΙΚΑΗΓΜΗΝΙΔΟΓΜΕΛΙΟ░, and does not note the ΟΓ below the line. This ΟΓ is the ending of the name in both lines 17 and 19.
18. The *Bulletin* reads ░Α░ΕΟΥΙΝΟΓΔΑΟΥ.
20. " " reads ░ΑΝΝΙΓΤΡΙΓ░ΥΡΟΥ.
21. " " reads ░ΡΙΕΠΕ░ΓΜΗΝΙΔ░, and fails to note the ΚΡΟΥ above the line.

```
        OCAICATTAΛOYNEOC
        KAXAPITWNNEAPXOY
        KΛCOYPN.OCCOYPNOYNEOC
   25   NEIKOΛAOCATTHΔOCΔICOCAEI
        MHNICNEAPKOYΛAΠOY
        ▨AYPHΛ·CTPATWNKWBEΛΛEOC
        ▨▨IICANTWNIOYKYNAKOC
        ▨▨▨COYPNOCMENANΔPOY
   30   ▨▨N▨ΛAX▨▨CΔICNEIKOΛ▨▨
        KACTWI▨▨▨ΔOYMHNIΔOC
        K▨NΛMI▨▨▨▨▨COYPNO▨
        ΔHMHCMHNIΔOCKIBYPOY
        OCAICATTHΔOCΔICTOYOCAEI
   35   MENANΔPOCΔICNEAPXOY
        ATTAΛOCΔICATTAΛOYTOYOC▨
```

Var. Lect.

22. The *Bulletin* indicates a break in fine, thus omitting the letters NEOC.

23. The *Bulletin* indicates as wholly defaced.

24.   "   "   indicates as wholly defaced.

25.   "   "   reads NIKOΛAOCATTH▨ΟΓΔIOCAEI.

26.   "   "   reads MHNICNEAPXOY▨▨▨.

27.   "   "   reads ▨▨YPPIACATWNIW▨▨▨.

28.   "   "   reads ANTWNIOCCYN▨▨▨.

29.   "   "   reads ▨▨▨ΘWNMENANΔPOY.

30.   "   "   reads ▨▨▨▨ICNEIKOΛ▨▨.

31.   "   "   reads ▨▨▨▨OYMENE▨▨.

32.   "   "   indicates as wholly defaced.

33.   "   "   reads ΔHMHCMHNIΔOCN▨▨.

34.   "   "   reads OCAICATTHΔOCΔICTW▨▨.

36.   "   "   ends the line with ΛOY, and does not indicate a break.

ΑΤΤΗϹΤΡΙϹΤΟΥΟϹΑΕΙ
ΟϹΑΗϹΔΙϹΜΗΝΙΔΟϹΟΚΕΡΗΓ▨
ΝΙΚΑΔΑϹΤΡΙϹΜΗΝΙΔΟϹ
40 ΜΗΝΙϹΔΙϹΟϹΑΙΝΕΟϹΡΗΓΕ
ΜΗΝΙϹΒΚΑϹΙΟΥΒΟΡΙϹΚΟΥ
▨ΩΒΕΛΛΙϹϹΤΡΑΤΩΝΟϹ

Var. Lect.

37. The *Bulletin* ends the line with ΟϹΑΕΙϹ, reading a superfluous C.
38. "   "   ends the line with ΜΗΝΙΔΟϹ▨.
39. "   "   ends the line with ΜΗΝΙΔ▨.
40. "   "   ends the line with ΝΕΟϹ, and does not indicate a break for the remaining letters.
41. The *Bulletin* reads ΜΗΝΙϹΒΙϹΑΕΙϹΥΒΟ▨ΙϹΚΟ.

After 41 the *Bulletin* inserts a wholly defaced line which does not exist.

42. The *Bulletin* reads ▨ΒΕΛ in init.

A.

['Αγαθῇ Τύχῃ · Ἔτους
. . . μηνὸς . . . Ὑπὲρ σω-]
τηρί[α]s ['Α]ννίας Φαυ[σ-]
τείνης καὶ δήμου 'Ο[ρ-]
μηλε[ω]ν ἐπὶ 'Αβασκά[ν-]
του πραγματευτοῦ·
5 Νεικόλαος Νεάρχου·
Νεικόλαος Συμμάχου·
Νεικάδας Ἡρακλείδου·
'Απολλόδοτος Μίλλακος·
Καλπούρνις Σοῦρνος·
10 Μῆνις Νεικάδου Ἡρ[α]κλείδου·
Μῆνις 'Οσαεί·
[Μῆν]ις 'Απολλωνίου·

Ἄτταλος Μουνδίωνος·
Ἡρακλείδης Νεικάδου·
15 Κρατερὸς Συμμάχου·
Ὀσαεὶς Ἀττάλου·
Μάρμας Μήνιδος·
Χάρης Μήνιδος Μάρμου·
Μῆνις τρὶς Μελίτων[ος]·
20 Νεικάδας δὶς Ἡρακλείδου·
Κάρπων Μήνιδος Οὖλο[υ]·
Κρατερὸς Μήνιδος Διδύμου·
Κάλλων Μήνιδος Μάρμου·
Νεικάδας Μήνιδος Νεικάδου·
25 Νεικάδας Ἡρακλείδου·
Μῆνις Καλαμίσκου·
Μῆνις Διομήδου·
Διονύσιος Νεικολά[ου]·
[Σύ]μμαχος Νεικολά[ου]·
30 [Μῆ]νις Μήνιδος Ὀσαεί·
Μῆνις Μήνιδος Νεικ[ά]δου·
Νεικάδας Νεικάδου Ἡρακλείδο[υ]·
[Ἀ]πολλόδοτος Νεικολάου·
Μόλυξ Κάστορος·
35 [Ἄτταλ?]ος Μεν[ά]νδρου·
[Ὀσαεὶ?]ς Χάρηδος Μό[λυκος?]

B.

Μῆνις Μήνιδος Οὐάλ[εντος]·
Κλαύδιος Κρατεροῦ·
Σύμ[μ]αχος Κρατεροῦ·
Ἀρίσ[τιπ]πος Ν[ε]ικολάου·
5 Διον[ύ]σιο[ς] Μήνιδ[ο]ς·
Κρατερὸς Ἐλίου·

Κάστωρ Μήνιδος ·
Διομήδης Μήνιδος ·
Μῆνις Χάρητος Ἡρακλείδο[υ] ·
10 Σοῦρνος δὶς νέος ·
Ἄτταλος Καλλικλῆδος ·
Ἄτταλος Δημητρίου ·
[Ἀ]βάσκαντος Ἀβασκάντου ·
Μῆνις Καλλικλῆδος ·
15 Λικίννιος δὶς νέος ·
[Μ]ένανδρος Κρατεροῦ ·
Μῆνις Συμμάχου ·
Ἄτταλος Ὀσαεί ·
Μῆνις Ἀττάλου [Δ]η[μ]ητρ[ί]ου ·
20 [Κ]αλικλῆς Καλικλλέους Μήνιδος ·
[Μ]ῆνις Μήνιδος Διομ[ή]δου ·
Μῆνις δὶς τοῦ Μίλακος ·
Γναῖος Μήνιδος ·
Κλ(αύδιος) Νέα[ρ]χος Μηνιανο[ῦ] ·
25 Μῆνις Νεικάδου Μήνιδος ·
Μῆνις Ἄττηδος ·
[Ἄ]ττ[α]λος δὶς Δημητρίου ·
[Μ]ένανδρος Νεάρχου ·
[Ἀρι]στέ[α]ς Μήνιδος Μίλακο[ς] ·
30 [Ἀ]ντώνιος δὶς Κύνακος ·
Κ(λαύδιος) Σοῦρνος τρίς ·
[Χ]άρης Γναίου Τ[έκ]των[ος?] ·
Σύμαχος Μάνου ·
[Ν]εικόλαος Συμμάχ[ου] ·
35 [Νε]ικάδας Σόλωνος Γρυπός ·
[Μ]ένανδρος Σούρ[νου] ·
[Σοῦρν]ος Μενάνδρο[υ] ·

C.

Μῆνις τρὶς Μήνιδος
. Νικάδου·
Δημήτριος Ἀττάλο[υ]·
Νικάδας Μήνιδος Νικάδου· .
5 Ἄττης δὶς τοῦ Ὀσαεί·
Κρατερὸς Κλανδίου·
Νεικάδας δὶς Μ[ή]νιδος·
Κ[λ](αύδιος) Μένανδρος Σούρνο[υ]·
Μῆνις δὶς Κυβύρου·
10 Μῆνὶς Μάρκου Μίλλακο[ς]·
Μῆνις τρὶς Οὐαδάρου·
Καλπούρ(νιος) Δάος Σούρνο[υ]·
Ὀσαεὶς Μήνιδος Οὐαδάρου·
Κλαῦδις δὶς Κρατεροῦ·
15 Δημήτριος Μήνιδος δὶς Νεικάδου·
Κιδράμας τρίς·
[Κ]αλλικλῆς Μήνιδος Μελίτωνος·
Καλ(πούρνιος) Σοῦρνος Δάου·
Μῆνις Καλλικλέους Μελίτωνος·
20 Μῆνις τρὶς Κιβύρου·
Ἀριστέας Μήνιδος τρὶς Μάκρου·
Ὀσαὶς Ἀττάλου νέος·
Κ[λ](αύδιος) Χαρίτων Νεάρχου·
Κλ(αύδιος) Σοῦρνος Σούρνου νέος·
25 Νεικόλαος Ἄττηδος δὶς Ὀσαεί·
Μῆνιϛ Νεάρκου Λάπου·
[Μ]. Αὐρηλ. Στράτων Κωβελλέος·
[Μῆν]ις Ἀντωνίου Κύνακος·
[Κλ.?] Σοῦρνος Μενάνδρου·
30 [Σύμμ]αχ[ος] δὶς Νεικολ[άου]·
Κάστω[ρ Νεικά]δου Μήνιδος·

K[αλά]μι[σκος δὶ]ς Σούρνο[υ·]
Δημῆς Μήνιδος Κιβύρου
Ὀσαὶς Ἀττηδος δὶς τοῦ Ὀσαεί·
35 Μένανδρος δὶς Νεάρχου·
Ἄτταλος δὶς Ἀττάλου τοῦ Ὀσ[αεί]·
Ἄττης τρὶς τοῦ Ὀσαεί·
Ὀσαὴς δὶς Μήνιδος ὁ κὲ Ῥήγ[ελλος]·
Νικάδας τρὶς Μήνιδος·
40 Μῆνις δὶς Ὀσαὶ νέος Ῥήγε[λλος]·
Μῆνις β′ Κασίου Βορίσκου·
[Κ]ωβέλλις Στράτωνος.

Lines of *C*, 38 and 40. The name ΡΗΓΕΛΛΟC is certain in No. 72, *A*, line 11.

Nos. 56–58.

Tefeny. Quadrangular stone serving as a foundation for the wooden pillar which supports the portico of the House of Mehcmet Bey. It is almost certainly inscribed on the fourth side also, but the stone cannot be removed without doing damage to the house. Its greatest present length is 0.97 m.; width, 0.56 m. Published in the Bulletin de Correspondance Hellénique, 1884, *p.* 497 *sqq.*

A.

The commencement of the lines were copied by A. H. Smith and verified by J. R. S. S. I was suffering from the fever at the time and could not bear to lie on my stomach with my head in the hole below me, consequently I have had to rely upon the impression for the body of the inscription on. this side. Fortunately it is good. Impression.[1]

[1] Ligatures occur in lines: 7, HNHB; 8, HN; 10, NK; 11, NHN, MH; 12, MN; 13, HM; 14, HM; 17, TH; 19, HN, HN; 20, HN, NMH; 24, HN; 26, MN, HMH, HP; 27, HP; 29, HN; 30, NMH, HNN; 31, NN.

```
     Δ̄
     ΙΟΝ
     ΜΟΥΝΟΙ
     ΣΕΙΣΧΣΤΙ
  5  ΑΚΟΤΗΤΑΦΥΣ
     ΞΑΘΛΑΗΖΕΙΣΚΑΙΔΩ
     \ΑΥΚΩΠΙΣΑΘΗΝΗΒΟΥ⌐
     )ΙΚΑΤΑΘΥΜΙΟΣΗΝΕΠΙb/
     \ΑΑΑ⋈Η⋈ΜΟΙΡΩΝ⋈ΤΕΣΣΑ
 10  ΕΙΠΤΩΝΚΑΙΜΟΥΝΟΙΤΕΣΣΑΡΕ_
     ΙΡΑΞΙΝΗΝΠΡΑΣΣΕΙΣΜΗΠΡΑΣ
     ΑΡΑΜΕΙΝΟΝΑΜΦΙΔΕΚΑΜΝΟι
     ΧΑΛΕΠΟΝΑΔΙΑΜΗΧΑΝΟΝΕΣΤι
     ΠΟΔΗΜΟΝΙΔΕΞΙΑΙΧΡΟΝΩΟΥΘ
 15  ΚΟΝΕΣΤΑΙ⋈ΓΓΑΑΑ⋈Θ⋈ΑΕΤΟΥΔΙΟ
     ΕΙΔΕΚΕΠΕΙΠΤΩΣΙΝΔΥΟΤΡΕΙΟΙΤΡ
     ΙΣΔΑΜΑΜΟΥΝΟΙ⋈ΑΕΤΟΣΥΥΙΠΕΤΗΣ
```

Var. Lect.

1. The *Bulletin* omits.
2. " " reads ΙΘΗ.
3. " " omits Ι at the end.
4. " " reads ΟΤ in fine.
5. " " reads ΦΥΙ in fine.
6. " " reads ΔΩϹ in fine.
7. " " omits \ in init.
8. " " reads ΘΥΜ and Β/ in fine.
9. " " omits \ in init.
10. " " reads ΡΕ in fine.
11. " " reads ꓶΒ in init., further on ΓΡΑ for ΠΡΑ.
12. " " ΝϹ in fine.
13. " " reads ΠΟΝΔΙΑ, and omits broken Ι at end.
14. " " reads ΠΟΛ in init., further on ΕΣ[Θ].
16. " " reads ΤΙ in fine.
17. " " reads Η⋗ in fine.

ΕΙ𝈝ΔΕ𝈝ΙΑΧΕΙΡΟ𝈝ΟΔΕΙΤΗ𝈝Ω̄ΝΕΠΙ

ΜΑΝΤΕΙΑΝΑΓΑΘΗΝ𝈝ΥΝΖΗΝΙΜΕΓΙ𝈝

20 ‾Ω̄ΤΕΥ𝈝ΗΕΦΗΝΟΡΜΑ𝈝ΠΡΑ𝈝ΙΝΜΗΘΕ

ΔΕΦΟΒΗΘΗ𝈝𝈝ΙΑΑΑΑ𝈝Ι𝈝ΔΑΙΜΟΝ

Ο𝈝ΜΕΓΙ𝈝ΤΟΥ𝈝𝈝ΕΙΤΗ𝈝ΜΟΥΝΟΙΤΕ𝈝

𝈝ΑΡΕ𝈝ΟΝΤΕ𝈝ΔΑΙΜΟΝΙΗΝΤΙΝΕΧΕΙ

ΕΥΧΗΝΑΠΟΔΟΝΤΙ𝈝ΟΙΕ𝈝ΤΑΙΒΕΛΤΕΙ

25 ΟΝΕΙΜΕΛΛΕΙ𝈝ΠΡΑ𝈝𝈝ΕΙΝΚΑΤΑΝΟΥΝΑ

ΝΕΡΙΜΝΑ𝈝ΔΗΜΗΤΗΡΓΑΡ𝈝ΟΙΚΑΙΖΕΥ𝈝

Ω̄ΤΗΡΕ𝈝Ε𝈝ΟΝΤΑΙ𝈝ΑΑΑΔΓ𝈝Ι𝈝ΤΥΧΗ

ΙΔΑΙΜΟΝΟ𝈝𝈝𝈝ΕΙΔΕΚΕΤΡΙ𝈝ΜΟΥΝΟ

ΤΕ𝈝𝈝ΑΡΑΤΡΙΑΟΠΕΜΠΤΟ𝈝𝈝ΤΗΝ

30 ͵𝈝ΙΝΜΗΠΠΡΑ𝈝Η𝈝ΗΝΝΥΝΕΠΙΒΑΛΛΗΙ

ΤΕΝΝΟΥ𝈝Ω̄ΕΟΝΤΑΘΕΟΙΚΑΤΕΧΟ

ΤΑΥΤΟΝΤΟΝΤΕΠΟΝΟΝΛΥ𝈝Ο`

ΟΙΚΑΙΟΥΘΕΝΚΑΚΟΝΕ𝈝ΤΑΙ

Λ𝈝ΙΑΝΕΙΚΗ𝈝ΕΙΔΕΚΕΤΡΕΙ𝈝

35 ΙΝΧΕΙΟΙΔΕΔΥΑΛΛΟ

ΛΥΗΔΑΑΘΕΛΕΙΣΤΑͻ

ͳϝΙΜΙΤΟΝ𝈝Ε‾

ΚΡΑΤΗ

ΜΙΟ

ΙΧ

Var. Lect.

17. The *Bulletin* reads ΕΙ𝈝Α𝈝ΔΙΑΧΕΙΡΟ𝈝 in init.
20. " " reads ‾Ω̄ΤΕ · 𝈝Η · ΦΗ in init.
21. " " reads ΟΝ in´fine.
22. " " reads 𝈝ΕΙΤ · 𝈝.
26. " " reads ΙΕΡ in init.
30. " " reads ΔΡΑ𝈝Η𝈝, and ΛΗ in fine.
31. " " reads ΤΕΝΝΟΥ in init.
32. " " reads 𝈝ΟΥ in fine.
34. " " reads Α𝈝ΙΑ in init.
36. " " reads ΓΥΗΛΑ in init. and ΤΑ in fine.
37. " " reads 𝈝Ε in fine. 40. The *Bulletin* omits.

B.[1]

This side was copied by W. M. Ramsay and copy verified
by J. R. S. S.   Copy and impression.

```
              - -
        Δ Ο ⸢ Τ Ι ⸢ Ο Ι Ε
        Δ Δ Α Α ⳧ Ι Δ ⳧ Α Γ ⸝
       Ο ⸢ ⳧ Ε Ι Δ Ε Κ Ε Π Ε Ι Π Τ Ω ⸢
       Ε ⸢ ⸢ Α Ρ Ε Ο Ι Κ Α Ι Δ Υ Ω Μ Ο Υ Ν C
      Ο Ν Ε Υ ⸢ Ε Ι ⸢ Ο Ι Δ Α Ι Μ Ω Ν Ο Δ Ο Ν
      Ι Ι Β Α Λ Λ Η Π Ε Ν Ψ Ε Ι Δ Ε Ι ⸢ Α Γ Α
      Ν ⸢ Ε Φ Ι Λ Ο Μ Μ Ε Ι Δ Η ⸢ Α Φ Ρ Ο Δ Ε Ι Τ Η
      ⸝ Ν Κ Α Ρ Π Ο Ι ⸢ Υ Π Α Γ Ε Κ Α Ι Α Π Η Μ Ο Ν Τ ⸝
      ⸝ Ο Ι Ρ Η ⳧ Α Γ Γ Δ Δ ⳧ Ι Ε Δ Ι Ο ⸢ ⸢ Ω Τ Η Ρ C
      Ε Ι ⸢ Μ Ο Υ Ν Ο ⸢ Δ Υ Ω Τ ⸢ Ι Ο Ι Δ Υ Ω Τ Ε Τ Ρ Ω Ο Ι
      Η Ν Ε Π Ι Β Α Λ Λ Η Π Ρ Α ⸢ Ι Ν Θ Α Ρ Ρ Ω Ν Ι Θ Ι Δ
      Ρ Α ⸢ Ε Ε Ν Χ Ε Ι Ρ Ε Ι Κ Α Λ Α Μ Α Ν Τ Ε Ι Α Θ Ε
      Ο Ι Τ Α Δ Ε Φ Η Ν Α Ν Μ Η Τ Ε Π Ι Ν Ο Υ Ν Α Λ Ε
      Ο Υ Ο Υ Θ Ε Ν Γ Α Ρ ⸢ Ο Ι Κ Α Κ Ο Ν Ε ⸢ Τ Α Ι
```

Var. Lect.

1.   The *Bulletin* omits.
3.   "   "   reads ΔΔ in init. and ΑΓΑΘ in fine.
5.   "   "   reads ΝΟ in fine.
6.   "   "   reads ΟΛ in fine.
7.   "   "   reads ΙΒΑ in init.
8.   "   "   omits Η in fine.
9.   "   "   omits vertical bar in init. and reads ΟΝ in fine.
10. "   "   reads ΘΙΡΗ in init. and ΡΟ in fine.
11. "   "   omits Ι in fine.
12. "   "   omits Δ in fine.
13. "   "   omits Ε in fine.

---

¹ Ligatures occur in lines: 8, MM; 9, NK, HM; 12, HN; 14, HN, NMH; 16,
MM; 17, HN; 18, HN; 26, NM; 29, NHM; 30, MH; 36, MHK; 37, HN; 38,
HNH.

ΑΑΑΙΙ⋈ΙΕ⋈ΔΙΟΣΑΜΜΩΝΟΣΜΟΥ
ΝΟΙΤΡΕΙΣΚΑΙΔΥΩ ‾ΕΙΤΑΙ⋈ΗΝΦΡΕΣΙ
ΝΟΡΜΑΙΜΕΙΣ‾　　ΥΤΗΝ(ΘΙΘΑΡ
‾ΩΝΠΑΝΔΕΣ　　　ΥΔΩΣΕΙΠΡΑ
20 ͘ΕΙΣΔΟΣΑΟ　　　‾ΥΥΙΒΡ.ΕΜΕ
ΤΗΣΣΩΤΗΡ　　　ΑΙ⋈ΓΓΓΓ
ΙΕ⋈ΤΥΧΗΣ　　　ΑΙΠΑΝΤΕΣ
‾ΡΕΙΟΙΗΔΕ　　　ΣΑΒΡΕΦΟΣ̵Ι
ΡΟΥΣΕΧΕΙΝ　　　ΑΠΑΛΙΝΒΛ∕
25 ‾ΤΗΣΕΚΑΙΑ　　　ΑΛΑΚΤΟΣΚΑ
ΟΤΕΕ̵ΕΙΣΙ　　　ΡΙΩΝΜΕΠΕΡΩ
‾ἈΣ⋈ΔΓΙΑΑ⋈　ΔΙΟΣ̵ΕΝΙΟΥ
ΤΕΤΡΩΕΙΣΚΑΙΤΡΙΟΣΚΑΙ̵ΕΙΘΟΣΕΙΣΚ∕
ΔΥΩΜΟΥΝΟΙΠΡΑ̵ΙΝΕΦΗΝΜΕΛΛΕΙΣΙ͘
30 ΝΑΙΜΗΣΣΠΕΥΔΟΥΠΩΓΑΡΟΚΑΙΡΟΣΚΑΙ
ΙΕΝΟΥΣΩΔΕΤΕΟΝΤΑΘΕΟΙΣΩΖΟˋ
‾ΤΟΙΜΩΣΚΑΙΤΟΝΕΝΑΛΛΗΧΩ
ΛΗ̵ΕΙΝΘΕΟΣΑΥΔΑ⋈ΙΓΓΓΑ
ΛΕΟΥΣ⋈̵ΕΙΤΟΣ̓ΕΙΣΚΑΙΤΡΕΙ

Var. Lect.

17. The *Bulletin* reads ΔΥΩ · ΕΙ, and omits Ι in fine.
18.   "   "   reads ΜΕΙΣ   Υ]ΤΗΝ.
19.   "   "   reads ΡΩΝΠΑΝΔ[ΕΕΣ?   Υ], and omits Α in fine.
20.   "   "   reads · ΕΙΣΔΟΣΑ[Β   ΥΥΙ.
21. . "   "   reads Ι̵Γ.
22.   "   "   reads · Ε in init.
23.   "   "   reads · ΡΕ in init. ˙and Σ̵̇ in fine.
24.   "   "   reads ΡΟΥ in init. and ΒΛ in fine.
25.   ". "   reads ΣΤΗ in init.
26.   ··   "   reads · ΟΤΕΕ̵ΕΙΣΚ in init. and Ρ̵ in fine.
27.   "   "   reads ΤΑ̸ΣΔΓΙΑΛ   ˙ΙΟΣ.
29.   "   "   reads ΔΥΩ in init. and ΣΙ in fine.
31.   "   "   reads ΖΟ in fine.

35

MOYNO≶EI≶⚹OYΠ♈KAI

EYΔEI≶ΔE≶YMHK≶NA

IΔ♈≶TI≶TEΛE♈NTYΦΛHN⁻

EIHNH≶YXABOYΛEYOYKA

⁻EMONEY≶EI⚹IΔΔAA⚹II

40　　　Θ⌒˂⁻　　　　⌐⌒ıΔYˀ

Var. Lect.

36. The *Bulletin* reads Θ]EY in init.
37. "　　"　　reads Δ♈≶ in init. and N⊏ in fine.
40. "　　"　　reads ΘO≶　　EOIΔY♈.

C.[1]

*This side was copied by J. R. S. S.; copy verified by
W. M. Ramsay. Copy and impression.*

IKAIE∠

A⋝INTAYTHNITP／

OKAIPO≶ENΓENE≶EI

.　NKAIOKINΔYNO≶ΠAP⟨

5　　KAIΠEPIT♈NAΛΛ♈NMAN

E≶TIKAΛ♈≶≶OI⚹AIΔΔΓ

EPAYNIOYMOYNO≶EI≶KA

!⁻O≶ΔY♈TETP♈OIKAITPIO≶

OYKE≶TINΠPA⋝ONTAKA

Var. Lect.

1. The *Bulletin* reads KAIE⌐.
2. "　　"　　reads A⋝ in init. and Nııı in fine.
4. "　　"　　reads TîAPA in fine.
8. "　　"　　reads O≶Δ in init.

---

[1] Ligatures occur in lines: 2, HN; 3, NΓ; 4, NK; 5, NM; 10, MHN; 11, HM;
13, HH, NH; 16, HN, HN, HN; 18, NM; 22, MH; 23, NMH; 24, HN; 25, HNH;
28, NM; 29, MH, NΓ; 30, HΓ; 35, NN; 37, HM; 39, HN; 42, HK.

10 ΙΛΓΝΩΜΗΝΑΜΕΡΙΜΝΑΣΟΥΤΕΓΑΡ
ΕΝΛΛΛΩΔΗΜΩΙΕΝΑΙΣΥΝΦΟΡΟΝ
ΕΣΤΙΝΟΥΤΩΝΟΥΜΕΝΟΣΑΙΣΘΗ
ΣΗΗΟΝΗΣΙΜΟΝΕΣΤΑΙ𐌇ΔΔΔΓΓ𐌇ΙΗ
ΔΑΙΜΟΝΟΣΙΚΕΣΙΟΥ𐌇ΤΕΤΡΩΟΙΤΡΙ

15 ΙΣΚΑΙΔΥΩΤΡΙΟΙ𐌇ΟΥΣΟΙΟΡΩΒΟΥ
\ΗΝΤΗΝΔΕΑΣΦΑΛΗΝΑΛΛΑΝΑΜΕΙ
ΟΝΕΥΠΡΑΞΕΙΣΕΣΤΑΙΣΕ
ΤΥΧΕΙΝΜΕΤΑΤΑΥΤΑΤΟΝΥΝΔΕΗ
ΞΥΧΟΣΗΣΟΘΕΟΙΣΠΕΙΘΟΥΚΑΙΕΠΕΙ

20 ΠΙΔΟΣΙΣΘΙ𐌇ΙΓΓΓ𐌇ΙΗ𐌇ΑΓΑΘΟ
ϹΡΟΝΟΥ𐌇ΖΕΙΘΟΣΚΑΙΤΕΣΣΑΡΕΣ
ΤΡΙΟΙ𐌇ΜΗΣΠΕΥΣΗΣΔΑΙΜΩΝΓΑΡ
ΑΝΘΙΣΤΑΤΑΙΑΛΛΥΠΟΜϹΙΝΟΝΜΗ
ΔΩΣΤΙΣΤΕΚΥΩΝΤΥΦΛΗΝΕΚΥΗΣΓ

25 ΛΟΧΕΙΗΝΗΣΥΧΑΒΟΥΛΕΥΟΥΚΑΙΣΟΙ
ΧΑΡΙΕΝΤΑΤΕΛΕΙΤΑΙ𐌇ΙΙΑΓΓ𐌇Ι·
ΕΛΠΙΔΟΣΑΓΑΘΗΣ𐌇ΕΥΟΔΑΣΟΙΠΑΙ
ΤΕΣΤΙΚΑΙΑΣΦΑΛΗΠΕΡΙΩΝΜΕΠΕ
ΡΩΤΑΣΜΗΔΕΦΟΒΟΥΔΑΙΜΩΝΓΑΡϹ

Var. Lect.

10. The *Bulletin* reads ΓΝΩ in init.
13. " " reads ΙΗ in fine.
14. " " does not indicate a break at the end.
16. " " omits / in init. and reads ΜΕΙ in fine.
17. " " reads ΙΟΝ in init. and ΣΕ in fine.
18. " " omits Τ in init.
19. " " reads ΣΥΧ in init. and ΕΠΕ in fine.
22. " " reads ΓΑΡ in fine.
23. " " reads ΥΠΟΜΕΙΝΟΝ.
24. " " reads ΥΗΣ in fine.
25. " " reads ΣΟ in fine.
29. " " reads ΓΑΡΟ in fine.

30 ⅃ΗΓΗΣΕΙΠΡΟΣΑΠΑΝΤΑΠΑΥΣΕΙ

    ‾ΑΡΛΥΠΗΣΧΑΛΕΠΗΣΛΥΣΕΙΔΥΠΟ

    ΝΟΙΑΝ⋈ΔΔΔΙΑ⋈ΙΘ⋈ΔΙΟΣΚΤΗ

    ΣΙΟΥ⋈ΘΑΡΣΩΝΕΝΧΕΙΡΕΙΚΑΙΕΦΕ

    ΛΠΙΔΟΣΕΣΤΙΝΟΧΡΗΣΜΟΣΩ⸗ΤΙ

35 ΜΑΝΥΕΙΚΑΙΤΟΝΝΟΣΕΟΝΤ/ .Ω

    'ΑΙΕΙΔΕΤΙΜΑΝΤΕΥΗΧΡ‾   ΧΡͰ

    ΕΙΣΑΠΟΛΗΜΥΗ⋈ΓΔΔΔ   Ρ

    ⅃ΟΥΚΕΡΔΑΕΝΠΟΡΟΥ⋈Μ

    ΒΟΥΛΗΝΣΑΙΣΙΦΡΕΣΙΝ

40 Ⅴ ΩΝΕΝΕΚΕΣΤΑΙΠΑΝΤΑ

    ΤΕΥΞΗΑΒΟΥΛΕΙΕΧΩΙ

    ΜΑΝΤΕΥΗΚΑΙΟΥΘΕΝ‹

    ΑΙ⋈ΙΓΓΓΑ⋈ΙΘ

        Λ

Var. Lect.

30.	The *Bulletin* reads ΔΗ in init.	
33.	" "	reads ΚΑΙΕ in fine.
34.	" "	reads ΩΣ in fine.
35.	" "	reads ΕΟΝΤ in fine.
36.	" "	reads ΕΙΕΙΔ in init. and ΥΗΧΡ in fine.
37.	" "	reads ΔΕΙΣΑΠΟΛΗΙΥΗ and omits Δ Ρ at
	the end.	
38.	The *Bulletin* reads ΙΟΥ in init., and omits Μ in fine.	
39.	" "	adds Ω in fine.
40.	" "	reads ΩΝ in fine.
41.	" "	reads ΕΙΕ · · · · Ι · · in fine.
44.	" "	reads ΛΙΤΕΙΛΝ.

δ

                 A.

ιον

   I. [ααααγ   ζ']   ✶✶✶✶✶

[εἰ δέ κε] μοῦνοι [τέσσαρες καὶ τρῶιο]ς εἷς,

[στι]				[κ]ακότητα φυσ . . .

[ς?] ἆθλα ἥξεις καὶ δω		γλ]αυκῶπις Ἀθήνη

βου[δορο?		ο]ι καταθ[ύμ]ιος ἦν ἐπι[βάλλῃ.]

II. [δα]ααα	η'	Μοιρῶν·

τέσσα[ρα δ' εἷς π]είπτων καὶ μοῦνοι τέσσαρε[ς ὄντες],.

[π]ρᾶξιν ἣν πράσσεις μὴ πράσ[σῃς, ἀλλ'] ἄρα μεῖνον·

ἀμφὶ δὲ κάμνο[υσιν], χαλεπὸν, ἀ[δ]ιαμήχανόν ἐστ[ι]

[ὑ]πὸ δῆμον ἰδέσ[θ]αι, χρόνῳ οὐθ[ὲν κα]κὸν ἔσται·

III. γγααα	θ'	Ἀετοῦ Διό[ς]·

εἰ δέ κε πείπτωσιν δύο τρεῖοι, τρ[ε]ῖς δ' ἅμα μοῦνοι,

ἀετὸς ὑψιπέτης εἰς δεξιὰ χειρὸς ὁδείτης,

ὧν ἐπὶ μαντείαν ἀγαθὴν σὺν Ζηνὶ μεγίσ[τ]ῳ

τεύξῃ· ἐφ' ἣν ὁρμᾷς πρᾶξιν μηθὲ[ν] δὲ φοβηθῇς·

IV. ζαααα	ι'	Δαίμο[ν]ος Μεγίστου·

[ξ]είτης, μοῦνοι τέσσαρες ὄντες,

δαίμονι ἥντιν' ἔχει[ς] εὐχὴν ἀποδόντι σοι ἔσται

βέλτειον εἰ μέλλεις πράσσειν κατὰ νοῦν ἃ [μ]εριμνᾷς·

Δημήτηρ γάρ σοι καὶ Ζεὺς [σ]ωτῆρες ἐσέσονται·

V. αααδγ	ι'	Τύχη[ς Εὐ]δαίμονος·

εἰ δέ κε τρὶς μοῦνο[ι, εἷς τέσσαρα, τρία ὁ πέμπτος,

τὴν [πρᾶ]ξιν μὴ πράξῃς ἣν νῦν ἐπιβάλλ[ῃ],

[καὶ] τ' ἐν νούσῳ ἐόντα θεοὶ κατέχο[υσί σε,] ταῦτον

τόν τε πόνον λύσο[υσί σ]οι καὶ οὐθὲν κακὸν ἔσται·

VI. [γγγαα]	ι[α']	Νείκης·

εἰ δέ κε τρεῖς [τρεῖοι εἰσ]ὶν, χεῖοι δὲ δύ' ἄλλο[ι],

		λήμ]ψῃ δ' ἃ θέλεις τα[

			[τειμι]τον σε[

[πάντα] κρατή[σεις]

B.

VII. [δ]δδαα	ιδ'	Ἀγ[αθοῦ Δαίμον]ος··

εἰ δέ κε πείπτωσ[ω τρεῖς τ]εσσάρεοι καὶ δύω μοῦν[οι],

[ἡγεμ]ονεύσει σοι δαίμων ὁδὸν [ἣν ἐπ]ιβάλλῃ
πένψει δ' εἰς ἀγά[πη]ν σε φιλομμειδὴς 'Αφροδείτη ·
[νῦ]ν καρποῖς ὕπαγε καὶ ἀπημον τ   οιρη
    VIII. αγγδδ   ιε΄   Διὸς Σωτῆρ[ος] ·
εἷς μοῦνος, δύω τρίοι, δύω τετρῶοι,
ἣν ἐπιβάλλῃ πρᾶξιν θαρρῶν ἴθι δρᾶσε ·
ἐνχείρει, καλὰ μαντεῖα θεοὶ τάδ' ἔφηναν,
μητ' ἐπὶ νοῦν ἀλέου · οὐθὲν γάρ σοι κακὸν ἔσται ·
    IX. ααασς   ιε΄   Διὸς Ἄμμωνος ·
μοῦνοι τρεῖς καὶ δύω [ξ]εῖται,
ἣν φρεσὶν ὁρμαίμεις [πρᾶξιν τα]ύτην ἴθι θαρ[ρ]ῶν
πανδε[σ   υ δώσει πρά[ξ]εις δός? 'Αθ[ήνη?]
ἠδὲ καὶ? Ζεὺς] ὑψιβρεμέτης σωτήρ [τε πατήρ τε?]
    X. γγγγγ   ιε΄   Τύχης
        αι πάντες [τ]ρεῖοι,
ἠδὲ   σα βρέφος ξ[ηρ]οὺς ἔχειν
        α πάλιν βλ[άσ]τησε καὶ ἀ   γ]άλακτος
κα[ί π]οτε ἔξεις ι   πε]ρὶ ὧν μ' ἐπερω[τ]ᾷς ·
    XI. δγσαα   [ιε΄]   Διὸς Ξενίου
τετρὼ εἷς καὶ τρῖος καὶ ξεῖθος εἷς κ[αὶ δ]ύω μοῦνοι,
πρᾶξιν ἐφ' ἣν μέλλεις ἰ[έν]αι μὴ ⟨σ⟩σπευδ', οὔπω γὰρ
                                        ὁ καιρός,
καὶ [τ'?] ἐ⟨ν⟩ νούσῳ δέ τ' ἐόντα θεοὶ σωζο[ῦσιν ἑ]τοίμως,
καὶ τὸν ἐν ἄλλῃ χώ[ρᾳ πόνον] λήξειν θεὸς αὐδᾷ ·
    XII. ςγγγα   [ις΄   Ἡρακλ]έους ·
ξεῖτος εἷς καὶ τρε[ῖς τρεῖοι], μοῦνος εἷς,
οὔπω καὶ[ρὸς, καθ]εύδεις δὲ σύ, μὴ κ(ε)νὰ [πράξῃς],
[μη]δ' ὡς τίς τε λέων τυφλὴν [ἐκύησε λοχ]είην ·
ἥσυχα βουλεύου κα[ί σοι θεὸς ἡγ]εμονεύσει.
    XIII. ςδδαα   ις΄ . . . .
[ξεῖ]θ[ος εἷς τεσσάρεοι] δύ[ω, καὶ δύω μοῦνοι] ·
. . .   .

C.

XIV. [ὅσσα   ιη′] . . . . .
[τέσσαρα δ' εἶς] καὶ ἑ[ξεῖται δύω καὶ δύω μοῦνοι],
[μὴ πράξῃς πρ]ᾶ[ξ]ιν ταύτην, [οὔπω γὰρ?] ὁ καιρός·
ἐν γενέσει   ν καὶ ὁ κίνδυνος παρ[αβαίνει],
καὶ περὶ τῶν ἄλλων μαν[τειῶν] ἐστι καλῶς σοι·

XV. ασδδγ   [ιη′   Διὸς Κ]εραυνίου·
μοῦνος εἶς κα]ὶ ξεῖτ]ος, δύω τετρῶοι καὶ τρῖος [εἶς,
οὐκ ἔστιν πράξοντα κα[τὰ] γνώμην ἃ μεριμνᾷς·
οὔτε γὰρ ἐν [ἄ]λλῳ δήμῳ ἰέναι σύνφορόν ἐστιν,
οὔτ' ὠνούμενος αἰσθήσῃ ᾗ ὀνήσιμον ἔσται.

XVI. δδδγγ   ι[η′]   Δαίμονος Ἱκεσίου·
τετρῶοι τρ[ε]ῖς καὶ δύω τρῖοι,
οὔ σοι ὁρῶ βού[λ]ην τήνδε ἀσφαλὴν, ἀλλ' ἀνάμε[ιν]ον·
εὐπράξεις ἔσται σ[ε] τυχεῖν μετὰ ταῦτα· τὸ νῦν δὲ
ἥ[σ]υχος ἧσο, θεοῖς πείθου, καὶ ἐπ' ἐ[λ]πίδος ἴσθι·

XVII. ϛγγγγ   ιη′ 'Αγαθο[ῦ Κ]ρόνου·
ξεῖθος καὶ τέσσαρες τρῖοι,
μὴ σπεύσῃς, δαίμων γὰ[ρ] ἀνθίσταται, ἀλλ' ὑπόμ[ε]ωον,
μη[δ]' ὡς τίς τε κύων τυφλὴν ἐκύησ[ε] λοχείην·
ἥσυχα βουλεύου, καί σο[ι] χαρίεντα τελεῖται·

XVIII. σσαγγ   ι[θ]′   'Ελπίδος 'Αγαθῆς·
εὔοδά σοι πά[ν]τ' ἐστὶ καὶ ἀσφαλῆ περὶ ὧν μ' ἐπερωτᾷς,
μηδὲ φοβοῦ· δαίμων γὰρ [ὁδ]ηγήσει πρὸς ἅπαντα·
παύσει [γ]ὰρ λύπης χαλεπῆς, λύσει δ' ὑπόνοιαν·

XIX. δδδσα   ιδ′   Διὸς Κτησίου·
θαρσῶν ἐνχείρει [κ]αὶ ἐφ' ἐλπίδος ἐστὶν ὁ χρησμός,
ω[ϛ?] τι μανύει; καὶ τὸν νοσέοντ' [ἀνασ]ώ[σ]αι·
εἰ δέ τι μαντεύῃ χρ   χρ   εἰς ἀπολήμψῃ·

XX. ·γδδδδ   [ιθ]   ['Ε]ρ[μ]οῦ Κερδενπόρου·
μ   βουλὴν σαῖσι φρεσὶν   ν]ων ἕνεκ' ἔσται,

πάντα [δέ σοι ἐπι]τεύξῃ ἃ βούλει ἔ[χ]ω[ν]. . . . .
μαντεύῃ καὶ οὐθέν [σοι κακὸν ἔστ]αι
XXI. · ςγγγδ   ιθ´.

I. 5. The letters BOYΖ seem to invalidate the reading of the *Bulletin*, βού[λεται πρᾶξίν σ]οι. The reading καταθύμιος is not certain.

II. 3. The sense seems to be for ἀλλ'] ἄρα μεῖνον in preference to οὐδ'] ἄρα μεῖνον of the *Bulletin*.

4. The reading ἀδιαμήχανον is reasonably certain. Διαμηχανάομαι means *to bring about, to contrive*. The Fates advise him, who has just consulted the oracle, to abstain from the matter he may have in mind, because among other things it is difficult to contrive and execute it.

5. ὑ[πὸ δῆμον is almost perfectly certain.

III. 3. εἰς δεξιὰ χειρὸς is plain and distinct on the impression.

IV. 2. [ξ]είτης is certain, so that the remarks in the *Bulletin*, loc. cit. p. 506, IV., do not hold good throughout.

V. 3. πράξῃς cannot be disputed, for although the π is slightly blurred on the impression, still the letter is certain.

4. κατέχο[υσί σε] seems more probable than κατέχο[υσί γε].

VI. 3. λήμ]ψῃ δ´ ἃ θέλεις must be read instead of ὑψηλὰ θέλεις.

VIII. 3. δρᾶσε is given by the *Bulletin* conjecturally. The Δ is not certain on the impression, but our copy gives the letter as certain (see *Bulletin*, p. 507, V.).

4. ἐνχείρει, imperative verb, must be read instead of ἐν χειρεὶ.

XIX. 2. ἐνχείρει, imperative verb, must be read instead of ἐν χειρεὶ, likewise ἐφ' ἐλπίδος instead of ἐπ' ἐλπίδος.

3. [ἀνασ]ώ[σ]αι seems to be demanded by the existing letters.

5. ἀπολήμψῃ must be read instead of ἀπολήψῃ. HM are in ligature.

For a similar inscription see *Papers of the American School*, Vol. III. Nos. 339-342. See also *Hermes*, X. p. 193 sqq.: *Rhein. Mus.*, VII. p. 251; Kaibel, *Epigr. Graeca*, p. 455.

No. 59.

Tefeny. Quadrangular cippus in the cemetery. Height, 1.75 *m.; within the mouldings,* 1.26 *m.; width,* 0.47 *m. Lines* 1–8 *copied by J. R. S. S., copy verified by W. M. Ramsay; lines* 9–19 *copied by W. M. Ramsay, copy verified by J. R. S. S.*[1]

```
   ΑΓΑΘΗΤΥΧΗ          ΕΤΟΥΣΔ
   
   ΤШΝΚΡΑΤΙϹΤШΝΤΕ
   ΚΝШΝΤΟΥΛΑΑΠΡΟΤΟΥΥΠΑΤΙΚΟΥ
   ΦΛΑΒΙΟΥΑΝΤΙΟΧΙΑΝΟΥΚΑΕΙϹΑ
 5 ΕΙΜΝΗϹΤΟΥΜΗΤΡΟϹΑΥΤШΝΠΟ
   ΠШΝΙΑϹΟΥΜΙΔΙΑϹΥΠΕΡϹШΤΗΡΙΑϹ
   ΑΥΤШΝΚΑΙϹШΤΗΡΙΑϹΔΗΜΟΥΟΡΜΗ
   ΛΕШΝ                           Λ
   ΑΥΡΜΗΝΙϹΒΚΑϹΙΟΥΒΟΡΙϹ
10 ΚΟΥΕΚΤШΝΙΔΙШΝΑΝΑΛШΜ
   ΤШΝΑΝΕϹΤΗϹΕΤΟΝΒШ
   ΜΟΝΤΟΙϹϹΥΝΙΕΡΕΙϹΙΝ
   ΤΟΥΔΙΟϹΕΓΕΝΕΤΑΔΑΠΑΝΗϹ✳
   ΓΑΙΟϹΝΙΓΡΟΥΒΑΓΑΝΔΕΥϹ
15 ΑΥΡΑΠΟΛΛШΝΙϹϹΤΡΑΤШΝ
   ΟϹΟΛΒΑϹΕΥϹΒΟΥΛΕΥΤΗϹ
   ΑΥΡΚШΒΕΛΛΙϹΔΙϹΙϹΤΡΑΤШΝΟ
   ΑΥΡΝΕΙΚΑΔΑϹΔΙϹΜΗΝΙΔΟϹ
   ΑΥΡΑΛΕΖΑΝΔΡΟϹΔΙϹΑΝΔΡΕ
```

[1] Lines 1 and 2 are on the moulding. In line 3, NT are in ligature. In line 4, a small Ι is inserted above and between the Ε and Ϲ. In line 5, the Η of ΝΗϹ is written above the Ϲ. In line 6, the reading ϹШΤΗΡΙΑϹ is certain, the stone-cutter having made an Μ by mistake instead of Η. At the close of line 7, a Λ has been incised below the Η; it can only belong to the name of the people, which in this instance must have two Λ Λ's.

'Αγαθῇ Τύχι · Ἔτους δ[ε?] . .
Τῶν κρατίστων τέ-
κνων τοῦ λανπρο(τά)του ὑπατικοῦ
Φλαβίου 'Αντιοχιανοῦ κα[ὶ τ]εῖς (= τῆς) ἀ-
5 [ε]ιμνήστου μητρὸς αὐτῶν Πο(μ-)
[π]ωνίας Οὐμιδίας ὑπὲρ σωτηρί[α]s
αὐτῶν καὶ σωτηρίας δήμου 'Ορμη⟨λ⟩
λέων ·
Αὐρ. Μῆνις β' Κασίου Βορίσ
10 κου ἐκ τῶν ἰδίων ἀναλωμ[ά-]
των ἀνέστησε τὸν βω-
μὸν τοῖς συνιερεῖσιν
τοῦ Διός · ἐγένετο δαπάνης ✳ . . . .
Γάϊος Νίγρου Βαγανδεύς,
15 Αὐρ. 'Απολλῶνις Στράτωνος 'Ολβασεὺς βουλευτής,
Αὐρ. Κωβέλλις δὶς 'Ιστράτωνο[s],
Αὐρ. Νεικάδας δὶς Μήνιδος,
Αὐρ. 'Αλέξανδρος δὶς 'Ανδρέ[α].

Line 9. We have before us the same person mentioned in No.
53 C, line 41.

The name 'Ιστράτων in line 17 is certain, but the Ι may be a
mistake of the stonecutter, see No. 53 C, line 42. If not, we might
assume that the natives of this region, like the Turks, felt it necessary
to insert an Ι before initial Σ, e.g. the Turkish *Ismir* [= Smyrna],
Isparta, Iskender, Istambul, etc. Compare No. 38 C, line 10, 'Ισκάλου.

Nos. 60–61.

*Tefeny. Theatre seat in the cemetery. Copied by W. M. R.
and J. R. S. S.* Bulletin de Correspondance Hellénique,
1878, *p.* 171.[1]

---

[1] In *B*, line 2, MH are in ligature.

A.

On the back of the seat.

ΕΤΟΥΣHΛΣΑΥΡΦΙΛΙ
ΡΟΣΔΗΜΗΤΡΙΟΥϺΙΚ
ΙΟΥΙΕΡΑΣΕΤΟ
ΕΤΟΥΣΣΛΣΑΥΡΠΑΠΗΣ
5 ΜΙΔΑΙΕΡΑΣΑΤΟΕΚΤϾΝ
▨ΑΙϾΝΘΕΛΙϾΣ

B.

On the right side of the seat.

ΑΠΟΛΛϾΝΙ
ΟϹΜΗΝΟϹ
ΜΕϹΑΝΒΡΙΟ▨
ΙΕΡΑΤΕΥϾΝ
5 ΜΗΝΙΤΟΛΗϹΕ
ϾΝΕΥΧΗΝ

Var. Lect.

2. The *Bulletin* reads ΟϹΜΗΝΙΔΟϹΤ, and in a footnote says:
   "ΜΗΝΙΔ lettres liées."

3. The *Bulletin* reads ΜΕϹΑΝΒΡΙΟ, and does not indicate a
   break in fine.

5. The *Bulletin* reads ΜΗΝΙΤΟΛΗϹΕ.

C.

On the left side we could make out nothing whatever, and the
copy as given in the *Bulletin de Correspondance Hellénique*, 1878,
p. 171, gave us no help, for here as elsewhere in this district the
copies of the *Bulletin* were constantly before us. Whereas we found
the inscription on the left side of the seat illegible, but that on the
back easy, the French gentlemen give a copy of the former and find
the latter hopeless.

A.

Ἔτους Ϛλσ' Αὐρ. Φίλι-
ρος Δημητρίου [Μ]ικ-
ίου ἱεράσετο.
Ἔτους ζλϚ' Αὐρ. Πάπης
5 Μίδα ἱεράσατο ἐκ τῶν
[ἰ]δίων θελίως.

B.

Ἀπολλώνι-
ος Μηνὸς
Μεσανβρίο[υ]
ἱερατεἰων
5 Μηνὶ Τολησέ-
ων εὐχήν.

A.

There can hardly be any doubt that the writer in the *Bulletin de Correspondance Hellénique* is right in considering that the era employed in this inscription and in others of this district is the Cibyratic, which has been fixed by M. Waddington (Le Bas-Waddington, *Voyage Archéologique*, 1213) to October, 25 A.D., not 25 B.C., as the writer in the *Bulletin de Correspondance Hellénique* (1878, p. 171 top) implies in his calculation.

Accordingly the dates given in this inscription, 236 and 237 (lines 1 and 4), correspond to the years 261 and 262 A.D.

Line 6. θελίως apparently stands for θελέως, *willingly.*

B.

" Apollonios, acting as priest of Men Mesanbrios, dedicates in discharge of a vow to Men of the Toleseis."

Two different gods Men are distinguished here; obviously they are the deities of two separate villages, one perhaps named Mesambria, and the other Tolesia (or possibly Todesia).

Nos. 62–63.

Tefeny. Stele in front of a house near that of Mehemet Bey. Copied by A. H. Smith.

A.

NACHΓYNHAYTOY
KAIMOYⳎAIOCKAI
IЄPΩNOIYIOIAY
TOYKAIMOYCAIOC
5 OANYYIOCAYTOY
APTЄMI▨▨▨TΩΠATPI
MNIACXAPIN

On the lower part of the same stele, but somewhat further around, is:

B.

ЄNACHΓYNHAYTOY
KAIMOYCAIOCKAI
IЄPΩNOIYIOIAY
TOYKAIMOYCAIOC
5 OȦNYYIOCAYTOY
APTЄMICIଠTଠΠATPI
MNIACXAPIN

A.

[Ἔ]νας ἡ γυνὴ αὐτοῦ ·
καὶ Μου[σ]αῖος καὶ
Ἱέρων οἱ υἱοὶ αὐ-
τοῦ καὶ Μουσαῖος
5 ὁ ἀνυψιὸς αὐτοῦ
Ἀρτεμι[σίῳ] τῷ πατρ[ὶ]
[μ]νίας χάριν.

B.

Ἔνας ἡ γυνὴ αὐτοῦ
καὶ Μουσαῖος καὶ
Ἱέρων οἱ υἱοὶ αὐ-
τοῦ καὶ Μουσαῖος
5 ὁ ἀνυψιὸς αὐτοῦ
Ἀρτεμισίῳ τῷ πατρὶ
μνίας χάριν.

No. 64.

Tefeny. Copied by W. M. Ramsay; copy verified by J. R. S. S.
Bulletin de Correspondance Hellénique, 1878, *p. 170.*

E T O Υ Ϛ T Υ Z

[Horseman]

Α Υ Ρ Δ Ι Ο Ν Υ Ϛ Ι Ο Ϛ

Δ Ι Ϛ Μ Ο Α Λ Ε Ι Δ Ο Ϛ

Θ Ε Ω Ϛ Ω Z Ο Ν Τ░░

Ε Υ Χ Η Ν ░

Var. Lect.

1. The *Bulletin* reads E T O Υ Ϛ T ⠀ Z I.
3. "⠀⠀⠀" ⠀⠀⠀reads M Ω Α Λ.
4. "⠀⠀⠀" ⠀⠀⠀reads N T I in fine.

Ἔτους τ[κ?]ζ΄
Αὐρ. Διονύσιος
δὶς Μοαλεῖδος
θεῷ Σώζοντι
εὐχήν.

If the conjecture K in line 1 be right, the inscription belongs to
the year 352 A.D.; see the commentary on Nos. 60–61.

Concerning Θεὸς Σώζων, or simply Σώζων, see *Bulletin de Corre-spondance Hellénique*, 1878, p. 171, No. 2 ; 172, No. 4, but especially 1880, p. 291 sq., where M. Collignon rejects the supposition that Σώζων represents *death* in the shape of a horseman, and thinks that the stones bearing reliefs of a horseman (along with inscriptions) are merely votive offerings dedicated to a God. It is noteworthy that Θεὸς Σώζων is mentioned on coins of Themissonion, a fact which tells strongly in favor of the close relation between this district and that in which Cibyra and Themissonion are situated.

No. 65.

Tefeny. On a rock near the tablet containing No. 68. The inscription is over a relief of Men, *who bears a club. Copied by W. M. Ramsay.*

MENEΛAOCMHNIΔOΓ
OPOΦYΛA
EYXHI
ETOYC
COP

Μενέλαος Μήνιδο[ς]
Ὀροφύλα[κι]
εὐχή[ν]·
Ἔτους
σο[β´]·

For Ὀροφύλαξ, see an inscription of Kara Agha, No. 165. The God is probably Men.

The date of the inscription is 297 A.D.

No. 66.

*Tefeny.   At the same place as No. 68, and on a similar
stone.   Copied by W. M. Ramsay.*

ΕΡΜΑΙΣΕΡΜΑΙΟΥ
ΜΑΣΑΔΙΤΨΑΔΕΛ
ΦΩ░░ΠΟΗΣΕΝΜΝΗ
░░░░░░░ΝΕΚΕΝ

'Ερμαῖς 'Ερμαίου
Μασᾶδι τῷ ἀδελ-
φῷ [ἐ]πόησεν μνή-
[μης ἕ]νεκεν.

No. 67.

*Tefeny.   Round column serving as a foundation stone to a
pillar in the house of Mehemet Bey.   Copied by A. H.
Smith; copy verified by J. R. S. S.*

ΔΗΜΗΤΡΙΟΣΔΗΜΗΤΡΙΟΥ
ΕΑΥΤΩΚΑΙΤΗΓΥΝΑΙΧΙ
ΖΩΝΕΠΟΗΣΕΝ

Δημήτριος Δημητρίου
ἑαυτῷ καὶ τῇ γυναιχὶ
ζῶν ἐπόησεν.

No. 68.

*Tefeny.   Tablet on a rock.   Copied by W. M. Ramsay
and A. H. Smith.*

E T O Y C B O P
I E P ϾΘ N B⅟⅞ K O I O Y

᾿Ετους βορ´
῾Ιέρων β´ Κοίου.

The date of the inscription is 197 A.D.

No. 69.

*Tefeny. Fragment in a wall. Copied by A. H. Smith;
copy verified by W. M. Ramsay.* Bulletin de Correspondance Hellénique, 1878, *p.* 264.

I O 'X' I X I
N E I Λ O C
E Λ E N O Y

It is broken only at the top; not at the sides as indicated in the
Bulletin.

. . . . .
Νεῖλος
῾Ελένου.

No. 70.

*Tefeny. Cippus with base in the cemetery. Copied by
W. M. Ramsay.*

K Λ Λ Y Δ I A
O C

Κλ[α]υδια[ν]ός.

No. 71.

Tefeny. On the street leading to Sazak. Copied by A. H. Smith and W. M. Ramsay. Bulletin de Correspondance Hellénique, 1878, *p.* 263.

░░░░░░░░░░░░░░░░
░░ K A I M H N ░░
I A Ϲ O N I K A I E I A

. . . καὶ Μήνι[δι]
Ἰάσονι καὶ Εἴᾳ.

Var. Lect.

1. The *Bulletin* reads ░KAIMH░.
2. " " reads ░AϹ in init. and EIΛ░ in fine.
3. " " reads ZΩϹIN, where the above copy does not give it.

Before I joined Messrs. Ramsay and Smith at Tefeny they had copied the following inscriptions at Hedje and Sazak.

Nos. 72-75.

Hedje. Partly in Bulletin de Correspondance Hellénique, *1878, p.* 255. C.I.G. 4367. *Impressions.*

A.

Copied by W. M. Ramsay; copy verified by A. H. Smith.[1]

---

[1] The N at the end of line 30 is regarded by Mr. Ramsay as somewhat doubtful. In line 8 the letter between Λ and Ω is certainly T, not Γ.

ΑΓΑΘΗ [uncut]

ΚΛΤΡΟΦΙΜΟCΙΤΑΛΙΚΟΥΕΤΙ
ΜΗCΕΤΟΝΟΧΛΟΝ✳Λ
ΓΑΕΙΟCΔΙCΜΗ [uncut]
5 ΑΝΕCΤΗCΕΝ
ΕΠΙΠΡΟΑΓΟΝΤΩΝΜΗΝΙΔΟCΔΙC
ΝΕΙΚΑΔΟΥ
ΛΤΤΑΛΟΥΟCΑΕΙΑΥΑΤΤΗC
ΔΙCΤΟΥΟCΑΕΙΠΡΟΑΤΩΝ
ΕΤΕΙΜΗCΕΝΤΟΝΟΧΛΟΝ
10 ✳ΡΟCΑΕΙCΜΗΝΙΔΟCΟCΑΕΙΟΥ
ΑΔΑΡΟΥΟΚΕΡΗΓΕΛΛΟCΕΤΙ
ΜΗCΕΝΤΟΝΟΧΛΟΝ✳Ν
ΚΑΛΛΙΚΛΗCΜΗΝΙΔΟCΜΕΛ
ΤΩΝΟCΕΤΙΜΗCΕΝΤΟΝΟ
15 ΧΛΟΝ☰C
CΟΛΩΝΝΙΚΑΔΟΥΜΕΝΕC
ΘΕΟCΕΤΙΜΗCΕΝΤΟΝ
ΟΧΛΟΝ✳Ν
ΠΑΝCΑCΚΛCΙΟΥΕΤΙ

Var. Lect.

The *Bulletin* starts out with the remark, "Cippe carré, écrit sur deux faces," whereas the stone bears legible inscriptions on all four sides.

1–11. The *Bulletin* omits these lines entirely.

12. The *Bulletin* reads ΟΝΟΧΛ.

13. " " reads ΚΑΙΚΛΗΜΗ.

14. " " reads ·· ΝΟCΕΤΙΜΗCΕ.

15. " " omits entirely.

16. " " reads ▨ΝΝΙΚΑΔΟΥΜΕ.

17. " " reads ▨ΟCΕΤ,ΜΗCΕΤΟΝ.

19. " " reads ΠΑΝΕΑCΚΛCΙΟΥΕΤΕ.

20 ΜΗϹΕΝΤΟΝΟΧΛΟΝΧΚΕ
ΜΗΝΙϹΝΕΑΡΚΟΥΛΑΠΟΥ
ΕΤΙΜΗϹΕΝΤΟΝΟΚΛΟΝ✳Ν
ΜΗΝΙϹΗΡΑΚΛΕΙΔΟΥΚΑϹ
ΤΟΡΟϹΕΤΕΙΜΗϹΕΤΟΝ
25 ΟΧΛΟΝ✳ΚΕ
ΑΤΤΑΛΟϹΚΕΜΑΡΚΟϹΟΙΔ
ΟΝΥϹΙΟΥΤΟΥΒΡΟΜΙΟΥΕ
ΤΙΜΗϹΑΝΤΟΝΟΚΛΟΝ✳▨
▨ΗΝΙϹΔΙϹΚΑΛΑΟΥΕ
30 ▨▨ΜΗϹΕΝΤΟΝΟΧΛΟΝ✳Ν
ΔΗΜΗϹϹΥΜΑΚΟΥΤΟΥΙΟΥ
ΝΙΟΥΕΤΕΙΜΗϹΕΝΤΟΝΟΚΛΟΝ
✳ΝΜΗΝΙϹΔΙϹ [uncut]
ΡШΝΟϹΕΤΕΙΜΗϹΕΝΤΟΝΟΚ
35 ΟΝ✳Κ·Ε

Var. Lect.

20. The *Bulletin* reads ✳ΚΕ in fine.
21. " " reads ΝΕΑΡΧΟΥ in the middle.
22. " " reads ✳Μ in fine.
23. " " reads ΚΑ▨ in fine.
26. " " reads ΑΤΤΑΛΟϹΚΕΧΛ · ΚΟΘΟ.
28. " " does not indicate a break in fine.
29. " " reads ▨ΝΙϹΔΙϹΚΑΛΑ · · · ·.
30. " " omits ✳Ν in fine, and does not indicate a break.
31. " " reads ΚΟΥ · · · · in fine.
32. " " reads ΟΧ▨ in fine.
33. " " reads ▨Ν · ΜΗΤΡΙΟΔ▨.
34. " " reads · ШΝ in init.
35. " " reads · ΟΝ in init.

B.

To the left of A. Copied by A. H. Smith; copy verified by W. M. Ramsay.

```
     KACIOCΔICTOYΠANCAETI
     MHCENTONOXΛON＊P
     KPATEPOC⫶⫶⫶AᴺΔIOYETIMH
     CENTONOXΛON＊P
 5   AΠO˙ΛΛOΔOTOCMHNIΔOC
     MIΔAKOCETIMHCENTONOX＊N
     MHNICTPICMEΛICCO    ΛON
     PΓOYETIMHCENTON
     OXΛON＊OEAΠOΛΛO
10   ΔOTOCΔICAΠOΛΛШ
     NIOYMIΛΛAKOCETEI
     MHCENTONOXΛON＊P
     MHNICAΠOΛΛOΔOTOY
     MIΛΛAKOCKEAYTOCE
15   TIMHCENTONOXΛON＊N
     MAPKOCMHNIΔOCΔIC
     CATAPAΔOCETIMHCEN
     TONOXΛON＊N
     ATTAΛOCM˙ENNEOYKIK
20   KOYETEIMHCENTONOKΛON＊N
     MENNE    ACKIKKOYETI
     MHCETONOKΛON＊N
     ⫶⫶⫶APAΓΔOYET
     ⫶⫶⫶HNIΔOC
25   ⫶⫶⫶YETEIM
     ⫶⫶⫶OXΛON
```

The *Bulletin* does not give this side at all, but cites under *A* three lines of Schönborn's bad copy. These lines correspond to lines 5–7 of the above copy.

C.

Copied by W. M. Ramsay; copy verified by A. H. Smith.

```
                                           ΑΡΚ
░░ΗΝΙCΜΕΝΑΝΔΡΟΥΜ
░░ΕΤΙΜΗCΕΝΤΟΝΟΧΛΟΝ✶ΚΕ
ΧΑΛΠΧΑΡΕΤωΝΝΕΑΡΚΟΥΜ░░
ΝΕΙΑΝΟΥΕΤΕΙΜΗCΕΝΤΟΝΟΚΛ
5  ΚΑCΤωΡΜΗΝΙΔΟ˙CΜΟ        ⋅
ΛΥΚΟCΕΤΙΜΗCΕΝΤΟΝΟ
ΧΛΟΝ✶ΡCΟΥΡΝΟCCΥΜ
ΜΑΧΟΥΚΡΑΤΕΡΟΥΕΤΕΙ
ΜΗCΕΝΤΟΝΟΧΛΟΝ✶Λ
10 ΑΝΤωΝΙΟCΜΗΝΙΔΟC
ΔΡϘϗΙΒΥΡΟΥΕΤΕΙΜΗCΕΝ
ΤΟΝΟΧΛΟΝ✶ΝΔΗΜΗCΜΗ
░░ΙΔΟCΚΙΒΥΡΟΥΕΤΕΙΜΗCΕΝ
░░ΟΝΟΧΛΟΝ✶ΚΕ
15 ░░ΗΝΙCΔΙΑCΚΟΥΡΙΔΟΥΒΙ
░░CΕΤΙΜΗCΕΝΤΟΝΟΧΛΟΝ✶░░
ΜΕΝΕCΘΕΥCΔΙCΦΥΡΡΟΥΕΤΙ
⋅ΜΗCΕΝΤΟΝΟΧΛΟΝΧΚΕ
░░ΑΔΑΥΑCΜΗ̧ΝΙΔΟCΚΑΔΑΟ
20 ░░ΑΙΟΥΙΟCΑΥΤΟΥΜ˙ΗΝΙCΚΑΔ
ΟΥΕΤΕΙΜΗCΕΝΤᵒΝΟΧΛΟΝ
ΟΝΗCΙΜΟCΜΗΝΙΔΟCΜΟΛΥ
ΚΟCΕΤ░░ΗCΕΝΤΟΝΟΧΛΟΝ✶Κ
░░ΠΠΟΛ░░░░░░CΟCΚΛΙ░░░
25 ░░ΙΟΓΑΥΤΟΥ░░░░░░░░░░░░
ΤΡΙCΕΤ░ΜΗCΑΝΤΟΝΟΧ
```

The *Bulletin* does not give this side at all.

D.

To the right of A. Copied by A. H. Smith; copy verified by W. M. Ramsay.

```
  ΜΗΝΙϹΑΧΙΛΛΕΟϹΕ
  ΤΙΜΗϹΕΝΤΟΝΟΧΛΟΝ*Ν
  ΑΧΙΛΛΕΥϹΜΗΝΙΔΟϹΜΟ
  ΥΝΓΟΥΕΤΙΜΗϹΕΝΤΟΝΟΧΛΟ*Ν
5 ΕΡΜΗϹΒΚΑΔΟΥΡΚΟΥΕΤΙ
  ΜΗϹΕΝΤΟΝΟΧΛΟΝ*Λ
  ΔΙΟΝΥϹΙΟϹΔΙϹΤΟΥΒΙΡШΝ
  ΟϹΕΤΙΜΗϹΕΝΤΟΝΟΧΛΟΝ*Κ
  ΜΕΝΝΕΑϹΔΙΟΝΥϹΙΟΥΜΕΝ
10 ΝΕΟΥΚΙΚΟΥΕΤΙΜΗϹΕΝΤΟ
  ΝΟΧΛΟΝ*ΛΕ
  ΔΗΜΟΦШΝΔΙΟΝΥϹΙΟΥΕΤΙΜ
  ϹΕΝΤΟΝΟΧΛΟΝ*ΚΕ
  ΜΕΝΝΕΑϹΚΑΡΠΟϹΑ
15 ΠΟΛΛШΝΕΙΟΥΕΙΕ
  ΡΕΟϹΕΤΕΙΜΗϹΕΝ
  ΤΟΝΟΧΛΟΝ*Ν
```

The *Bulletin* does not give this side at all.

A.

Ἀγαθῇ [Τύχῃ. Ἔτους . . .?]
Κλ(αύδιος) Τρόφιμος Ἰταλικοῦ ἐτίμησε τὸν ὄχλον * (τριάκοντα)·
Γάειος δὶς Μή(νιδος)
5 ἀνέστησεν
ἐπὶ προαγόντων Μήνιδος δὶς Νεικάδου,
[Ἀ]ττάλου Ὀσαεί· Αὐ(ρ). Ἄττης
δὶς τοῦ Ὀσαεὶ προά[γ]ων

ἐτείμησεν τὸν ὄχλον
10 ✳ (ἑκατὸν) · Ὀσαεὶς Μήνιδος Ὀσαεὶ Οὐ-
αδάρου ὁ κὲ Ῥήγελλος ἐτί-
[μ]ησεν τὸν ὄχλον ✳ (πεντήκοντα) ·
Καλλικλῆς Μήνιδος Μελ[ί]-
τωνος ἐτίμησεν τὸν ὄ-
15 χλον ✳ (διακόσια) ·
[Σ]όλων Νικάδου Μενεσ-
θέος ἐτίμησεν τὸν
ὄχλον ✳ (πεντήκοντα) ·
Πάνσας Κ[α]σίου ἐτί-
20 μησεν τὸν ὄχλον [✳] (εἴκοσι καὶ πέντε) ·
Μῆνις Νεάρκου Λάπου
ἐτίμησεν τὸν ὄχλον ✳ (πεντήκοντα) ·
Μῆνις Ἡρακλείδου Κάσ-
τορος ἐτείμησε τὸν
25 ὄχλον ✳ (εἴκοσι καὶ πέντε) ·
Ἄτταλος κὲ Μάρκος οἱ Δ[ι-]
ονυσίου τοῦ Βρομίου ἐ-
τίμησαν τὸν ὄκλον ✳ . . .
[Μ]ῆνις δὶς Κα[δ]άου ἐ-
30 [τί]μησεν τὸν ὄχλον ✳ (πεντήκοντα ·
Δημῆς Συμάκου [τοῦ Ἰου-]
νίου ἐτείμησεν τὸν ὄκλον
✳ (πεντήκοντα) · Μῆνις δὶς [Δά-]
ρωνος ἐτείμησεν τὸν ὄκλ-
35 ον ✳ (εἴκοσι καὶ πέντε) ·

B.

Κάσιος δὶς τοῦ Πάνσα ἐτί-
μησεν τὸν ὄχλον ✳ (ἑκατόν) ·
Κρατερὸς [Κλ]α[ν]δ[ί]ου ἐτίμη-

σεν τὸν ὄχλον *(ἑκατόν)·

5 Ἀπολλόδοτος Μήνιδος
Μί[λ]ακος ἐτίμησεν τὸν ὄχλον *(πεντήκοντα)·
Μῆνις τρὶς Μελισσό-
ργου ἐτίμησεν τὸν
ὄχλον *(ἐβδομήκοντα καὶ πέντε)· Ἀπολλό-
10 δοτος δὶς Ἀπολλω-
νίου Μίλλακος ἐτεί-
μησεν τὸν ὄχλον *(ἑκατόν)·
Μῆνις Ἀπολλοδότου
Μίλλακος κὲ αὐτὸς ἐ-
15 τίμησεν τὸν ὄχλον *(πεντήκοντα)·
Μάρκος Μήνιδος δὶς
Σατάραδος ἐτίμησεν
τὸν ὄχλον *(πεντήκοντα)·
Ἄτταλος Μεννέου Κίκ-
20 κου ἐτείμησεν τὸν ὄκλον *(πεντήκοντα)·
Μεννέας Κίκκου ἐτί-
μη[σε τὸν ὄ]κλον *(πεντήκοντα)·
[Μῆνις? Σμ]αράγδου ἐτ[ίμησε κ.τ.λ.]
[Ὁ δεῖνα Μ]ήνιδος
25 [Διονυσιο?]ῦ ἐτίμ-
[ησεν τὸν] ὄχλον·

C.

[Μ]ῆνις Μενάνδρου Μάρκ[ο-]
[υ] ἐτίμησεν τὸν ὄχλον *(εἴκοσι καὶ πέντε)·
[Κ]αλπ(ούρνιος) Χαρέτων Νεάρκου Μ[η-]
νειανοῦ ἐτείμησεν τὸν ὄκλ[ον κ.τ.λ.]
5 Κάστωρ Μήνιδος Μό-
λυκος ἐτίμησὲν τὸν ὄ-
χλον *(ἑκατόν)· Σοῦρνος Συμ-

μάχου Κρατεροῦ ἐτεί-
μησεν τὸν ὄχλον ✳ (τριάκοντα)·
10 Ἀντώνιος Μήνιδος
[δὶς Κ]ιβύρου ἐτείμησεν
τὸν ὄχλον ✳ (πεντήκοντα)· Δημῆς Μή-
[ν]ιδος Κιβύρου ἐτείμησεν
[τ]ὸν ὄχλον ✳ (εἴκοσι καὶ πέντε)·
15 [Μ]ῆνις Διασκουρίδου Βί-
[τυο]ς ἐτίμησεν τὸν ὄχλον ✳ . . . .
Μενεσθεὺς δὶς Φύρρου ἐτί-
μησεν τὸν ὄχλον [✳](εἴκοσι καὶ πέντε)·
[Κ]αδαύας Μ[ή]νιδος Καδάο[υ]
20 [κ]αὶ ὁ υἱὸς αὐτοῦ Μῆνις Καδ[αύ-]
ου ἐτείμησεν τὸν ὄχλον·
Ὀνήσιμος Μήνιδος Μόλυ-
κος ἐτ[είμ]ησεν τὸν ὄχλον ✳ (εἴκοσι)·
[Ἀ]ππολ[λώνιος . . . .] σος κ[α]ὶ [ὁ]
25 [υ]ἱὸς αὐτοῦ [ὁ δεῖνα]
τρὶς ἐτ[ί]μησαν τὸν ὄχ[λον κ.τ.λ.]

D.

Μῆνις Ἀχιλλέος ἐ-
τίμησεν τὸν ὄχλον ✳ (πεντήκοντα)·
Ἀχιλλεὺς Μήνιδος Μο-
ύνγου ἐτίμησεν τὸν ὄχλο(ν) ✳ (πεντήκοντα)·
5 Ἑρμῆς β΄ Καδούρκου ἐτί-
μησεν τὸν ὄχλον ✳ (τριάκοντα)·
Διονύσιος δὶς τοῦ Βίρων-
ος ἐτίμησεν τὸν ὄχλον ✳ (εἴκοσι)·
Μεννέας Διονυσίου Μεν-
10 νέου Κίκου ἐτίμησεν τὸ-
ν ὄχλον ✳ (τριάκοντα καὶ πέντε)·

Δημοφῶν Διονυσίου ἐτίμ[η-]
σεν τὸν ὄχλον ✳(εἴκοσι καὶ πέντε)·
Μεννέας Κάρπος Ἀ-
15 πολλωνείου Εἱε-
ρέος ἐτείμησεν
τὸν ὄχλον ✳(πεντήκοντα).

No. 76.

Hcdje. On a red column. Copied by *W. M. Ramsay.*

O I C Θ E Ϭ N
A Y T O K P A
Ϲ E Π T I M
Π E P T I N A K
A Δ I A B
K A I A Y T O
Y P H Λ I Ϭ

B Ϲ I Λ E Ϭ N   I O Y Λ I A Ϲ
M H T P I⬛⬛⬛Ϲ⬛⬛⬛
          Λ I I O K I B Y

[Τ]οῖς θεῶν [ἐπιφανεστάτοις]
Αὐτοκρά[τορι Καίσαρι Λουκίῳ]
Σεπτιμ[ίῳ Σεουήρῳ Εὐσεβεῖ]
Περτίνακ[ι Σεβαστῷ Ἀραβικῷ]
5 Ἀδιαβ[ηνικῷ Παρθικῷ Μεγίστῳ]
καὶ Αὐτο[κράτορι Καίσαρι Μάρκῳ]
[Α]ὐρηλίῳ [Ἀντωνείνῳ Εὐσεβεῖ]
[καὶ Ποπλίῳ Σεπτιμίῳ Γέτᾳ

. . . . μεγά-
10   λων] Βασιλέων [καὶ] Ἰουλίᾳ Σ-
     [εβαστῇ] μητρὶ Κά]σ[τρων].
     [ἀπ]ὸ Κιβύ[ρας]

. . . . . .

Mr. Ramsay makes a note that below BY in line 12 there was perhaps MK, but that the reading is exceedingly doubtful, and he would have looked rather for IΘ.

For a similar inscription of Cibyra, see *Bulletin de Correspondance Hellénique*, 1878, p. 597.

No. 77.

Hedje. Upper part of a Stele in the cemetery. Copied by A. H. Smith.

M H N I C Δ O Y Λ Π A
E Π O H C E M H N I
Δ Ι T W A Δ E Λ Π W
K A T H M H T P I K A I A Y
T W K A I T H Γ

Μῆνις Δού[δ]α?
ἐπόησε Μήνι-
δι τῷ ἀδελπῷ
κα(ὶ) τῇ μητρὶ καὶ αὐ-
τῷ καὶ τῇ γ[υναικὶ]
[αὐτοῦ μνίας χάριν].

No. 78.

Sazak. Copied by W. M. Ramsay and A. H. Smith. Partially in Bulletin de Correspondance Hellénique, 1878, *pp.* 173–4.[1]

ΠΟΚΟΙΤΗΣ·Μ·ΚΛΛ
ΠΟΥΡΝΙΟΥΛΟΓΓΟΥ
ΠΑΤΡΩΝΟΣΙΔΙΟΥ
[A bust]
Μ·ΚΑΛΠΟΥΡΝΙΟΣ
5 ΕΠΙΝΕΙΚΟΣΜΙΣΘΩ
ΤΗΣΤΩΝΠΕΡΙΑΛΑΣΤοΝ⅞
ΤΟΠΩΝΔΙΙΜΕΓΙΣΤΩ

Var. Lect.

1. The *Bulletin* reads ΟΙΟΙ   ΚΛΛ.
2.   "   "   reads ΟΙΤΟ.
3.   "   "   reads ΟΧ.
6.   "   "   reads ΤΗΣΤΩΝΤΕΡΙΔΑΣΤΟΝ.

['Α]πὸ κοίτης Μ. Κ[α]λ-
πουρνίου Λόγγου
πάτρωνος ἰδίου
Μ. Καλπούρνιος
'Επίνεικος μισθω-
τὴς τῶν περὶ 'Αλαστο[ν]
τόπων Διὶ Μεγίστῳ.

June 11. Tefeny, *via* Kayalü and Mandja, to Kaldjik, 4 h. 42 m. We travel for the most part in the plain, but pass around some low hills which always remain on our left.

---

[1] Ligatures occur: line 1, ΤΗΣ; line 5, ΝΕ; line 6, ΗΣ, ΝΠΕ; line 7, ΜΕ.

No. 79.

Kayalii. Quadrangular cippus.

ҧΤΕΜΩΝΜ·ΚΑΛΠΟΥΡΛ
ΟΥΛΟΝΓΟΥΔΟΥΛΟ.ϹΟΙ
ΚΟΝΟΜΟϹΔΥΟΝΥϹΩΘΕ
ΩΕΠΗΚΩΕΥΧΗΝ

['Αρ]τέμων Μ. Καλπουρ[νί-]
ου Λόγγου δοῦλος οἰ-
κονόμος Δυονύσῳ Θε-
ῷ 'Επηκ(ό)ῳ εὐχήν.

Concerning 'Επήκοος, see Le Bas-Waddington, *Voyage Archéologique,*
1173; *C.I.G.* 4900–4902 ; *Bulletin de Correspondance Hellénique,*
1878, p. 173, No. 5 ; 1879, p. 336, No. 5.   See also the next inscription.
M. Καλπούρνιος Λόγγος is mentioned in No. 78.

No. 80.

*Kaldjik. Quadrangular Stele in the house of Halil Bey.
Copied by J. R. S. S.; copy verified by W. M. R. and
A. H. Smith.*

ΚѠΒΕΛΛΙϹΔΙϹ
ΤΟΥΑΤΤΗ
ΠΟϹΕΙΔѠΝΙ
ΕΠΗΚΟѠ
ΕΥΧΗΝ

Κωβέλλις δὶς
τοῦ Ἄττη
Ποσειδῶνι
'Επηκόῳ
εὐχήν.

Ποσειδῶν Ἐπήκοος is mentioned in an inscription of *Karamanli* published in the *Bulletin de Correspondance Hellénique*, 1878, p. 173, No. 5.

No. 81.

Kaldjik. Badly defaced inscription belonging to a ruined mausoleum in the plain below the village.

M H N I C Δ I C M E N A N Δ P O Y⅚ K A I K A K I E Π Σ Ψ I Λ I▓

H Γ Y N H A Y T O Y C P E I C Δ H M H T P O Σ L▓

A Y T O I Σ K A I Ξ ͞ ͞ ͞A▓ K K K A͞ ͞ ͞I N▓T▓Y▓

[uncut]

κ H I ε I ▓⁗▓M H N I Δ O Σ Δ A O Y P I▓O▓

[uncut]

T ⲱ N Π P O▓N O N T ⲱ N▓

Μῆνις δὶς Μενάνδρου καὶ . . . .
ἡ γυνὴ αὐτοῦ (ἱ)[ε]ρεῖς Δήμητρος . . . . [ἑ-]
αυτοῖς κα[τεσκεύασ]α[ν? καὶ] κα[θιέρωσαν? τὸ μνημεῖον]
. . . . Μήνιδος Δάου Ῥι[ζ]ο[ῦντος?]
. . . . . των προ[γό]ν[ω]ν των-.

No. 82.

Kaldjik. Round basis in the house of Halil Bey. Copied by J. R. S. S.; copy verified by W. M. Ramsay and A. H. Smith.[1]

▓E M M E N I Δ H Σ
▓Λ P X O N T O Σ
▓Λ Π Y K E Y Σ
▓E M I Δ I
▓

---

[1] The second letter in line 3 is somewhat doubtful. It might possibly be a T, but we all agreed to write it as given in the uncial text.

.... Ἐμμενίδης
.... [ἄ]ρχοντος
[. . . α]πυκεὺς
['Αρτ]έμιδι
[εὐχήν].

No. 83.

Kaldjik. Copied by J. R. S. S., W. M. R., A. H. S.

```
▨▨▨▨  ⎡         ▨▨▨▨
ΠΟΛ   ⎢ ornament ⎤ ШΝΙᏟ▨
ϹΔΙϹ  ⎣         ⎦ ΠΛΟ▨
ΥΤШΝΙΕΤ[Ι▨
ΟΛΝΤΟΛΟΙ
ΙΟΝΕΥΚΗΝ
```

.... ['Α-]
πολωνι[ο-]
ς δὶς Πλο-
ύτωνι ....

. . . .

.... εὐκήν.

At Kaldjik I bade a final farewell to Messrs. Ramsay and Smith. The general plan of my journey made it impossible for me to work longer in concert with them.

June 12. Kaldjik, *via* Bademli, Mussalar, Eïnesh, to Hadjilar, 5 h. 30 m. We traverse an open, rolling country along the north-western edge of the valley of the Gebren Tchai, a district blank on the old map, but which contains a number of villages.

No. 84.

Mussalar.   Quadrangular cippus.   Copy.

ONHCIMOCTAT
AKA'TOΛMINA
HΓYNHAYTOYB
ШMONANECTHC
ANHMHCXAPIℵ

'Ονήσιμος Τατ-
ᾶ καὶ Τολμῶνα
ἡ γυνὴ αὐτοῦ ⟨τὸν⟩ β-
ωμὸν ἀνέστησ-
αν ⟨μ⟩νήμης χάρι[ν].

Nearly one hour east of Eïnesh is the site of an ancient town, now wholly deserted.   The remains are not unworthy of notice.   Among other things may be mentioned the tombs, most of which are round buildings, with massive stone foundations.   These were probably ἐξέδραι or ψαλίδες.   I have met with them also at Isaura Vetus and at Anabura (cf. *Papers of the American School*, Vol. III. No. 187 and p. 203; also Nos. 339–342.   Mr. Ramsay identifies this site with "Palaiopolis, or Alieros, the latter being the native name" (cf. *American Journal of Archaeology*, Vol. III. p. 161).

No. 85.

Hadjilar.   In the cemetery.   Copy.

AYPHΛΛI
OCΔHMHC
NANAΔOC
ZШNEAYTШ
KAITHΓYNE
KIAYTOYAΠI
ANH

Αὐρή⟨λ⟩λι-
ος Δημῆς
Νάναδος
ζῶν ἑαυτῷ
καὶ τῇ γυνε-
κὶ αὐτοῦ ᾿Απι-
ανῇ.

June 13. Hadjilar to Buldur, 3 h. 38 m.

No. 86.

Buldur.   The inscription is in a panel on a fluted column
in the court of a house.   Copy.

```
▨▨▨▨(ΛΛΟϹΑΝΤΙ
ΟΧΟΥϹΚΡΑΓΟΥΟΛ
ΥΝΠΙΑϹΓΥΝΗΚΑΙ
ΝΕШΝΥΕΙΟϹΑΝΕΘ
5      ΗΚΑΝ
ΑΝΤΙΟΧΟϹΥΙΟϹ
```

[῎Αττα]λος ᾿Αντι-
όχου Σ[κ]ράγου, ᾿Ολ-
υνπιὰς γυνὴ, καὶ
Νέων νεὶος ἀνέθ-
5   ηκαν
᾿Αντίοχος υἱός.

The names Antiochus and Attalus? give an approximate date to
the inscription.   Note the Nom. in lines 1–4 instead of the Acc.

June 14.   Buldur to Isparta, 4 h. 55 m.   We pass Kyshla, Eski
Yer, and Tcharshü ; at Tcharshü we begin the ascent of the mountain,

and in 1 h. 14 m. the watershed is reached. The road descends
through a narrow gorge for 35 m., when the western edge of the plain
of Isparta is reached. This plain was explored by me in 1885 ; see
Papers of the American School at Athens, pp. 326–351.

No. 87.

*Isparta (Baris). The inscription is on a highly ornamented
lintel of a door or gateway, possibly the door of a church.*
Bulletin de Correspondance Hellénique, 1879, *p.* 343, *No.*
20. *Copy.*

ΥΠΕΡΜΝΗΜΗ ΕΚΑΙΑΝΑΠΑΥϹΕωϹΠΑΥΛΟΥΕΠΙ Κ ᔆΔΙΟΥΑΝΕ
ΚΤΙϟΕΝΤΟΝ ΝΑΟΝΤω ΝΑΡΧΑΝΓΕΛωΝΙΝΔᔆΓΕ

Ὑπὲρ μνήμης καὶ ἀναπαύσεως Παύλου Ἐπίκ[τα ? (or
Ἐπικ[τήτου?] or Ἐπικ[ύδους?]) Δίου ἀνέκτισεν τὸν ναὸν
τῶν ἀρχαγγέλων ἰνδ(ικτιῶνος) [σ]ε'.

Line 1. The *Bulletin* reads ΜΗϹΚ, ΕΠΙΚΥΔ, but the *siglum
interpunctionis* is plain.
Line 2 fin. The *Bulletin* reads ϟ Γ▨.

No. 88.

Isparta. Epistyle block in the pavement of the street. Copy.[1]

ΘΥΒΡΥΩΝΙ'ΟΞΤΕ
ΤΟΝΚΛΕΙΝΟΝΚΗΡΨϹΕ

Θυβρύων? . . . . .
τὸν κλεινὸν Κηρυ . . .?

---

[1] In line 2, NK are in ligature.

No. 89.

Isparta.   In a fountain.   Letters wet and blurred.   Copy.[1]

```
⟨⟩ Ῡ Ϲ Υ Ν Ε Ρ ˢ Κ Α Ι Χ Ρ Ι Ϲ ˥ ˢ Κ Α Ι Α Γ Ι Ο Υ ⫿ Ν
Τ Ο Υ Α Γ Ι Ο Υ Γ Ε Ο Ρ Γ Ι Ο Υ Ϲ Υ Ν Ο ⧖
Δ Ι Α Ε Ϲ Τ Υ Α ⋈ Ν Ω Ν Ε Ι Ω Α Ν Ν Η Ϲ
Π Ρ Ε Ϲ Β Υ Τ Ε Ρ Ο Ϲ Ε Ρ ˢ Α Β Ρ Α Μ Ι Ϲ
5  Τ░ Ω Α Γ Ω Ν Α Ν Τ Ω Ν Ι Ϲ
Ε Ι Ω Α Ν Ν Η Ϲ Α Ρ Τ Ε Μ Ω Ν
Μ Λ Ι Κ Ι Ϲ Ζ Ω Τ Ι Κ Ο Ϲ Φ Ι Λ Ι Π Π Ο
Κ Λ Η Μ Ε Ν Τ░ Υ Ρ Ι Α Κ Ο Ϲ
Θ ⸽ϲ⸽ Ο Δ Ο Υ ▨ [water-spout] ▨ Ρ Τ Ε Μ Ω Ν
10  Μ Α Ρ Τ Υ Ρ Ι Ϲ░ Ω Τ Ι Κ Ο Ϲ
Α Λ Ε Ζ Α Ν Δ Ρ Ο Ϲ Δ Ι Μ Ι Τ Ρ Ι Ο Ϲ
Α Τ Τ Α Λ Ο Ϲ Ζ Ω Τ Ι Κ Ο Ϲ
Φ Ι Λ Ι Π Π Ο Ϲ Π Α Τ ░░
```

[Ἔτο]υς νν΄ ἔρ(γον) καὶ Χρισ[τοῦ] καὶ ἁγίου . . . . .
τοῦ ἁγίου Γεωργίου συνο-
δία . . . . . . Εἰωάννης
πρεσβύτερος ἐρ(γεπιστήσας) Ἀβράμις
5 [προ]άγων Ἀντῶνις
Εἰωάννης Ἀρτέμων
Μ[α]ῗκις Ζωτικὸς Φίλιππο[ς]
Κλήμεντ[ος Κ]υριακὸς
Θ[ε]οδούλου Ἀ]ρτέμων
10 Μαρτύρις [Ζ]ωτικὸς
Ἀλέξανδρος Διμίτριος
Ἄτταλος Ζωτικὸς
Φίλιππος Πατ[ρίκιος?]

---

[1] Ligatures occur: line 1, NE; line 3, NE, NHϹ.

For a similar inscription in Baïyat (Seleucia Sidera) in the plain of Isparta, see *Papers of the American School at Athens*, Vol. III. No. 465.

Προάγων is the title of an official in Ormele ; see the inscriptions of Karamanlü and Tefeny given above (Nos. 41 A ; 43).

If the restoration of line 1 be correct, then the date of the inscription is 450 A.D.

No. 90.

Isparta. In a corner of a house by the above fountain.
Copy.

X P I Ξ T E
B O H Θ I

Χριστὲ βοήθι.

No. 91.

Isparta. In the court of a Medressi. The inscription is on the left side of a stone with a shell-like niche, in which possibly once stood a statuette. Copy.

P O Δ Ω N Ϲ E Λ E Y
K O Y T P Ω I Λ O Y
Ἱ E P A Ϲ A M E N O Ϲ %
T O N K A Θ H Ϲ E M O
N A E P M H N E K T Ω N
I Δ I Ω N I Δ P Y Ϲ A T O

'Ρόδων Σελεύ-
κου Τρωΐλου
ἱερασάμενος
τὸν καθησέμο-
να? Ἑρμῆν ἐκ τῶν
ἰδίων ἱδρύσατο.

June 15. Isparta to Egherdir, 5 h. 29 m. Leaving Isparta we traverse the ˙plain whose topographical features are described in the *Papers of the American School,* Vol. III. p. 332. Near Güle Önü we head about east. Forty-five minutes east of Egherdir the watershed is reached. A sharp descent brings us to Egherdir, situated near the southwestern end of the lake which bears its name. Here I was shown two ancient steelyards. The four sides of the bronze beams were all different, each side being apparently intended for a different standard of weight. The great intervals corresponding to our one, two, three, etc., pound notches, were marked by letters of the Greek alphabet. From the style of the letters the steelyards must be placed in the late Roman or early Byzantine period. The heavy weight was a bronze head of Zeus, filled with lead. The workmanship of this head was much too good for the period mentioned. I could not buy them.

June 16. Egherdir to Gelendos, 7 h. 58 m. In 42 m. we cross the Boghaz Su by a bridge at its exit from Egherdir Göl. It is a strong, deep, and very rapid stream. I afterwards traced it up in 1885 (cf. *Papers of the American School,* Vol. III. pp. 309–310, and pp. 317–318). The road henceforth for four hours is very difficult. It follows in general the coast of the lake, and crosses a succession of spurs or benches of the mountain, which fall off more or less perpendicularly into the lake. A new road has been constructed recently : much blasting has been done, and abutments have been built where the road lies along the edge of the lake. In 4 h. 42 m. from Egherdir we reach the Devrend, which is situated just at the point where the rough road over the Demir Kapu (called also Eyerim Bel) ceases. Henceforward the road lies in a plain. We pass a large Seldjuk Khan 50 m. north of the Devrend.

From this point on the reader may consult the large map in Vol. III. of the *Papers of the American School.*

June 17. Gelendos to Yalowadj, 5 h. 14 m. For the topographical details of this region of country, and for numerous inscriptions not given in this present volume, see the *Papers of the American School,* Vol. III. pp. 218–278.

Nos. 92-93.

Yalowadj (Antiochia Pisidiae). In the cemetery near the mill opposite Hissar, immediately on the road leading from Yalowadj to Ak Shehir, and thirty minutes distant from Yalowadj. Quadrangular cippus; inscription badly defaced. Copy.[1]

A.

```
AYPΔIONYCI
ONTONAΞIO
ΛΟΓΩΤΑΤΟΝΕ
ΚΑΤΟΝΤΑΡΧΟΝ
5  ΡΕΓΕΩΝΑΡΙΟΝ
ΗΛΑΜΠΡΑΤΩΝΑΝ
ΤΙΟΧΕΩΝΜΗΤΡΟ
ΠΟΛΙϹΕΠΕΙΚΙΑϹ
ΤΕΚΛΙΤΗΕΕΙΡΗ
10  ΝΗϹΕΝΕΚΑ        ·
```

B.

```
ΤΟΝΔΕϹΕΜΥ
ΓΔΟΝΙΗΔΙΟΝΥ
ϹΙΟΝΑΝΙ▨▨▨
ΠΟ▨▨▨▨▨
5  ΚΑΙΤΗϹΕΙΡΗΝΗϹ
ΕΤΕΜΜΑ
```

A.

Αὐρ(ήλιον) Διονύσι-
ον τὸν ἀξιο-

---

[1] I have a note to the effect that in *A*, line 5, ΤΕΓΕ might be read. In *A*, line 9, ΤΗΕ for ΤΗϹ is certain. In *B*, line 6, the reading is certain. Inscription *A* was published in uncials in my *Preliminary Report*, etc., p. 9.

λογώτατον ἑ-
κατόνταρχον
5 [λ]εγεωνάριον
ἡ λαμπρὰ τῶν Ἀν-
τιοχέων μητρό-
πολις ἐπ(ι)εικίας
τε κ[α]ὶ τῆ(ς) εἰρή-
10 νης ἕνεκα.

B.

Τόνδε . . . .
. . . . . . Διονύ-
σιον Ἀν[τιοχέων?]
πό[λις ἐπιεικίας τε]
καὶ τῆς εἰρήνης
[ἕνεκα].

No. 94.

*Yalowadj. In the wall of a house opposite a Djami. Once
an inscription of eighteen lines, but all except the first two
and the last line have been dug out.*[1] C.I.L. *III.* 301 ; *Le
Bas-Waddington*, Voyage Archéologique, *III.* 1825. *Copy.*

A N T I O C H
A E C A E S A R E
[Fifteen lines missing.]
S A C E R A V G

---

[1] The excavated part is fully two inches deep; some one evidently intended to
make a trough of the stone.

No. 95.

Yalowadj. In the Djami of Kizildje Mahallii. Copy.[1]

▨▨▨KIANOᴄ
▨▨▨ᴋΑΙΛΟΓΙᴄΤΗᴄ
▨▨▨ΝΤΙΟΧΕWΝ
▨▨▨ΟΠΟΛΕWᴄ
▨▨▨ΟᴄΤΑΓΙΟΝ

[Λου]κιανὸς
[. . . . κα]ὶ λογιστὴς
[τῆς τῶν 'Α]ντιοχέων
[μητρ]οπόλεως
. . . .

Concerning the functions of the λογιστὴς (curator urbis), see Marquardt, *Römische Staatsverwaltung*, I. p. 162 sqq.; Henzen in *Annali dell' Instituto*, 1851, pp. 5, 16, 17; *Revue Archéologique*, 1863, VII. p. 373, and the commentary on p. 377; Franz, *Fünf Inschriften und fünf Städte in Kleinasien*, pp. 15–18; *Bulletin de Correspondance Hellénique*, 1884, p. 389, No. 8; 1885, p. 395, and the commentary on p. 396; 1886, p. 222, No. 4; 1878, p. 523; *Mittheilungen des Deutschen Archaeologischen Institutes in Athen*, 1878, p. 56, No. 1; *Journal of Hellenic Studies*, VI. p. 348; Μουσεῖον καὶ Βιβλιοθήκη τῆς Εὐαγγελικῆς Σχολῆς, 1875, p. 118, No. 17; 1878, p. 29, Nos. 230, 231, p. 33, 237; 1885, p. 76, No. 484.

No. 96.

Yalowadj. Quadrangular cippus in the corner of a house opposite the barracks. Copy.

ΗΒΟΥΛΗ	Ἡ βουλὴ
Τ▨Ν	τ[ὸ]ν
ᴄΕΚΟΥΝΔΟΝ	Σεκοῦνδον
ΕΠΙΤΗ	ἐπὶ τῇ
ᴄΤΡΑΤΗΓΙΑ	στρατηγίᾳ.

---

[1] Ligatures occur in lines 2, Hᴄ; 3, WN; 4, Wᴄ.

No. 97.

Hissar, a village half an hour east of Yalowadj: grand tablet, whose length is 1.27 m.; width, 0.65 m.   Copy.[1]

TYXHNEY
MENHTH
KOΛωΝεl
ATIBEPIO
ΠΟΛΕΙΤωΝΠΑΠ
ΗΝωΝΟΡΟΝΔΕ
ωΝΒΟΥΛΗΔΗΜΟC

Τύχην εὐ-
μενῆ τῇ
Κολωνεί-
ᾳ Τιβεριο-
πολειτῶν Παπ-
ηνῶν Ὀρονδέ-
ων βουλὴ δῆμος.

No. 98.

Yalowadj.   In the wall of a house opposite the Djami nearest the barracks.   Length, 1.7 *m.; width,* 0.52 *m.*   C.I.L. *III.* 291 ; *Le Bas-Waddington,* Voyage Archéologique ; *Henzen,* Inscr. Lat. Selectarum Collectio, 6912, *with a note on p.* 521, *all from a copy of Hamilton.   Copy and impression.*

---

[1] Published in uncials in my *Preliminary Report,* p. 13; afterwards in the *American Journal of Archaeology,* 1885, p. 143.

P·F·STEL·SOSTٍ🖤

TI·FETIALI·LEG·AVG

PRO·PR·PROVINC·GAL

PISID·PHRYG·LVC·ISAVR·

PAPHLAG·PONTI·GALAٍ

PONTI·POLEMONIANI

ARM·LEG·LEG·X̄ĪĪĪ·GE Nٍ

DONAT·DON·MILITARIB

EXPEDIT·SVEBIC·ET·SARM

COR·MVR·COR·VALL·COR

AVR·HAST·PVR·TRIB·VE

XILL·TRIB·CVRAT·COLO

NIOR·ET·MVNICIPIOR·PRAE

FRVM·DAND·EX·S·C·PRAETOR

AED·CVRVL·Q·CRET·ET·Cٍ

·TRIB·LEG·XXIII·PRIMIGEN

IIIVIR·A·A·A·FF·

THIASVS·LIB

· · · · · · · · · · · ·

P(ublii) f(ilio), Stel(latina), So[spi-]

ti, fetiali, leg(ato) Aug(usti)

pro pr(aetore) provinc(iarum) Gal(atiae),

Pisid(iae), Phryg(iae), Lyc(aoniae), Isaur(iae),

Paphlag(oniae), Ponti Gala[t](ici),

Ponti Polemoniani,

Arm(eniae), leg(ato) leg(ionis) XIII Ge[m](inae)·

donat(o) don(is) militarib(us)

expedit(ione) Suebic(a) et Sarm(atica)

cor(ona) mur(ali), cor(ona) vall(ari), cor(ona)

aur(ea), hast(is) pur(is) trib(us), ve-

xill(is) trib(us), curat(ori) colo-

nior(um) et municipior(um), prae(fecto)

frum(enti) dand(i) ex S(enatus) c(onsulto), praetor(i),

    aed(ili) curul(i), q(uaestori) Cret(ae) et C[yr](enarum),
    trib(uno) leg(ionis) XXIII Primigen(iae),
    triumvir(o) a(eri) a(rgento) a(uro) f(lando) f(eriundo)
    Thiasus lib(ertus).

My copy and impression justify the restorations of Henzen, but not Borghesi's conjecture of SOLLERTI in lines 1 and 2. While in the presence of the stone I made a note to the effect that the end of line 1 must be either SODI or SOSI, and now an inspection of the impression convinces me of the accuracy of my note. The letter immediately following SO cannot possibly be an L, and while a D might not be absolutely impossible, yet all the indications go to show that the letter must be an S. The letters of line 1 are larger than those of the following lines, and unfortunately no D occurs in it; but nevertheless it is safe to assert that the fragmentary letter in question is too narrow for a D, whereas it is of exactly the same size as the corresponding part of the two S's in line 1 and resembles them in every way.

    The CAPPADOCIAE looked for by Mommsen in line 3 is certainly wanting: evidently the administrations of Cappadocia and Galatia were separate at this time.

No. 99.

Yalowadj. In western cemetery. Ephemeris Epigraphica, 1884, *p.* 576, *No.* 1344. *Copy and impression.*[1]

P · A N I C I O
P · F · S E R · M A X I
· M O · P R A E F E C T O
C N D O M I T I · A H E N o B A R
5 B I · P · P · L E G XII F V L M · P R A E F

---

[1] Ligatures occur: line 4, HE; line 5, AE; line 13, VM. For a discussion of the inscription in its historical bearings, see the *Ephemeris Epigraphica* as cited.

```
CASTRORLEGΠAVGIN
BRITANNIAPRAEFEXER
CITVQVIESTINAEGVPTO
DONATO·AB·IMP·DONIS·
```
10
```
MILITARIBVS·OB·EXPEDI
TIONEM·HONORATO·
CORONA·MVRALI·ET·
HASTA·PVRA·OB·BELLVM
BRITANNIC[uncut]CIVITAS
```
15
```
ALEXANDR·QVAEEST
INAEGVPTO    H·C·
```

P(ublio) Anicio,

P(ublii) f(ilio), Ser(gia), Maxi-

mo, praefecto

Cn. Domiti Ahenobar-

5 bi, p(rimo)p(ilo) leg(ionis) XII Fulm(inatae), praef(ecto)

castror(um) leg(ionis) II Aug(ustae) in

Britannia, praef(ecto) exer-

citu(i) qui est in Aegypto,

donato ab imp(eratore) donis

10 militaribus, ob expedi-

tionem honorato

corona murali et

hasta pura ob bellum

Britannic(um), civitas

15 Alexandr(ia) quae est

in Aegypto h(onoris) c(ausa).

Line 2 fin. *EE* omits I.

Line 4. *EE* reads NB.

Line 7 fin. *EE* omits ER.

Line 10 fin. *EE* reads DITI.

Line 12. *EE* reads CORON, omitting A.

No. 100.

Yalowadj. In the western cemetery. Length, 1.56 m.; width, 0.56 m. Ephemeris Epigraphica, 1884, *p.* 579, *No.* 1345. *Copy and impression.*[1]

```
      LEG·AVG·PROPR·PROVINC·GALAT·PHRYG
      PISID·LYCAON·PAPHLAG·ITEM·ADCENSVS·PA
      PHLAG·LEG·LEG·T·M·P·FINGERMINFER·PRAETO
  ·   RI·CANDIDATO·IMPTRAIANIAVG·GERM·DA
      CICI·PARTH·TRIBPLEB·CANDID·EIVSDIA▨   ·
      VIR·SAC·FAC·ADACTA·SENATVS·Q▨
      EQ·R·TRIB·LEG·X̄XĪI·PPF·X̄VIR·STLĪ▨
      [uncut]        ET   [uncut]
```

leg(ato) Aug(usti) pro pr(aetore) provinc(iarum) Galat(iae), Phyg(iae),
Pisid(iae), Lycaon(iae), Paphlag(oniae), item adcensus Pa-
phlag(oniae), leg(ato) leg(ionis) I M(inerviae) P(iae) F(idelis) in Germ(ania) Infer(iore), praetori candidato imp(eratoris) Traiani Aug(usti) Germ(anici) Da-
cici Parth(ici), trib(uno) pleb(is) candid(ato) [e]iusd[em XV
vir(o) sac(ris) fac(iundis), ad acta Senatus, qu[aest]ori', [VI vir(o)]
eq(uitum) R(omanorum), trib(uno) leg(ionis) XXII P(rimigeniae) P(iae) F(idelis), xvir(o) stli[t](ibus) [iudi-]
[candis]

---

[1] Ligatures occur: line I, PHR, YG. Only such dots are given as are certain on the stone.

No. 101.

Yalowadj. *In the wall of the Djami inside the town nearest the barracks.* C.I.L. *III.* 295 ; *Le Bas - Waddington,* Voyage Archéologique, *III.* 1818; *Henzen,* Inscr. Lat. Sel. Collectio, 6157, *from a copy of Hamilton.* *Copy and impression.*[1]

```
   C · A L B V C I O C · F·
   S E R · F I R M O A E D
   ii V I R · Q V I P E C V N I
   A M D E S I I N A V I T P E I·
5  T E S T A M E N T V M A I
   C E R T A M E N G Y M N I C V I
   Q V O    A N N I    A C I E N
   D V M D I E B V S F E S T I S
        L V N A E
10 D           D
```

C(aio) Albucio, C(aii) f(ilio),
Ser(gia), Firmo, aed(ili),
duumvir(o), qui pecuni-
a[m] des[t]i[n]avit pe[r]
5 testamentum a[d]
certamen gymnicu[m]
quo[t]anni[s f]acien-
dum diebus festis
lunae.
10 D(ecreto) D(ecurionum).

In line 4 the stone, my copy and impression do not bear out Hamilton's DESIGNAVIT. Following the DES are four vertical strokes, between the last two of which the stone has been battered, but still the diagonal bar of an N is reasonably certain. It is not

---

[1] In line 6, MNI are in ligature.

impossible that the stonecutter failed to cross his T, and that DESTINAVIT is the true reading.

The last lines of the inscription are omitted in Hamilton's copy.

No. 102.

Yalowadj. In the wall of the Djami near the barracks. Length, 1.38 *m.; width,* 0.65 *m.* C.I.L. *III.* 292; *Le Bas-Waddington,* Voyage Archéologique, 1817; Ephemeris Epigraphica, 1885, *p.* 575, *No.* 1340. *Copy and impression.*[1]

```
C·NOVIO·C·NOVI
PRISCICOS·ETFLAVONIAE
MENODORAE·FIL·SER·RVS
TICO·VENVL·APRONIANO
5  XVIR·STLITIBIVDICANDIS
TRIRI.ATICLLEGVIFERCAPAR
```

C. Novio, C. Novi
Prisci co(n)s(ulis) et Flavoniae
Menodorae fil(io), Ser(gia), Rustico Venul(eio) Aproniano
5. (decem)vir(o) stlitib(us) iudicandis
[trib](uno) [l]atic[l](avio) [l]eg(ionis) VI Fer(atae) . . .

I do not understand CAPAR in line 6; compare CAPARC in No. 103.

Line 1, fin. *EE* reads NO𝑉.
Line 2, fin. *EE* reads N𝐼.
Line 3, fin. *EE* reads R𝑉.
Line 4, fin. *EE* reads NC.
Line 6, *EE* reads TRIBᴵATICᴵᴵEGVIFERCAPA.

[1] Ligatures occur in lines: 2, NI (= Ṅ); 4, NI (= Ṅ); 5, ND.

No. 103.

Yalowadj. In the corner of a house opposite a fountain.
Copy.

```
APROI
XVIR·STLIT·IV
TRIBL·G·LEG··VII
  CAPARC·QVAES
  CAND·LEG·ASI
  TRIB·CAND
     DESIC
  VICD
```

[C. Novio, C. Novi
Prisci co(n)s(ulis) et Flavoniae
Menodorae f(ilio), Ser(gia),
Rustico Venuleio]
Apro[niano]
(decem)vir(o) stlit(ibus) iu(dicandis) ,
trib(uno) l(ati[c]lavio) leg(ionis) VI [Fer(atae)]
caparc . . . .? quaes[tori]
cand(idato) leg(. . . .) Asi(ae)
trib(uno) cand(idato)
        desi[gnato]
[vic(us)] D . . . . . .

No. 104.

Hissar. In a Djami. Copy.

```
C·NOVIO·C·
NOVI·PRISCI
COS·ET·ΓLAVON
```

```
  MENODORAE
5 FIL·SER·R·STICO
  ⸜ENVL·APRON
```

C(aio) Novio, C(aii)
Novi Prisci
co(n)s(ulis) et [F]lavon[iae]
Menodorae
5   fil(io), Ser(gia), R[u]stico
[V]enuleio Apron[iano,
(decem)vir(o) stlitib(us) iudicandis,
trib(uno) laticl(avio) leg(ionis) VI ferr(atae)].

No. 105.

*Kuyudjak, about two hours southeast of Yalowadj. Block:
length, 1.08 m.; width, 0.57 m. My copy verified by J. H.
Haynes. Copy.*

```
      LCORNELIO
      LFSERMARCEL
      LOAEDQGRAM
      MATI ⅡVIRO
5     HORTENSIA·M·
      FGAILLA·AVVN
      CVLOSVOOPTI
      MOETAMANTIS
      SIMOOBMERI
10    TAEIVS
```

L(ucio) Cornelio,
L(ucii) f(ilio), Ser(gia), Marcel-
lo, aed(ili), q(uaestori), grammati, (duum)viro

5 Hortensia, M(arci)
f(ilia), Gaïlla avun-
culo suo opti-
mo et amantis-
simo ob meri-
10   ta eius.

Lines 3-4, *Grammati* is Greek; dative of γραμματεύς.
For *Gaïlla*, see No. 106.

No. 106.

*Yalowadj. Quadrangular Stele in the court of the house
of Isa Oghlu. Copy.*

ΠΟΥΠΙΛΛΙΑ
CΑΤΟΥΡΝΙΝΑ
ΚΑΙΠΟΥΠΙΛΛΙ
ΟCCΑΡΙΝΟC
5 ΕΖΑΔΕΛΦΗ
ΓΑΙΛΛΗΓΛΥΚΥΤΑΤΗ
ΙΔΙΑΜΝΗΜΗC
ΧΑΡΙΝ⚹

Πουπιλλία
Σατουρνῖνα
καὶ Πουπίλλι-
ος ['Ε]άρινος
5 ἐξαδέλφῃ
Γαΐλλῃ γλυκυτάτῃ
ἰδίᾳ μνήμης
χάριν.

Line 6 is in very small letters, and was inserted as an afterthought
after the inscription had been engraved. The inscription has been
inserted here because of the name Gaïlla, which occurs in No. 105.

No. 107.

Gemen [*Yemen*].   *Large cubical stone lying under a tree by the brook which runs through the gardens.   Copy and impression by J. H. Haynes and J. R. S. S.   Copy.*[1]

```
 ///CARISTANVSHAMYRVS
 CCARISTANIVSAGAPETVS
 CCARISTANIVSHAPIVS
 CCARISTANIVSFAVSTVS
 CCARISTANIVSVALENS
 CCARISTANIVSPOTHVS
 CCARISTANIVSFLACCVS
 CCARISTANIVSFELIX
```

[C](aius) Caristan(i)us Hamyrus
C(aius) Caristanius Agapetus
C(aius) Caristanius Hap(t)us
C(aius) Caristanius Faustus
C(aius) Caristanius Valens
C(aius) Caristanius Po[t]hus
C(aius) Caristanius Flaccus
C(aius) Caristanius Felix.

See the note to No. 108.

No. 108.

Yalowadj.   In the pavement by a canal.   Copy.

```
 ////////////////////////////////
 ////ΑΝΛΟΥΚΙϹ////////
 ΓΑΤΕΡΑΠΑΥΛΛΑ
 ΓΥΝΑΙΚΑΓΑΙΟΥΚΑ
 ΡΙ Ϲ ΤΑΝΙΟΥΦΡΟΝ
```

[1] In line 1 there is no I between N and V as in the other lines.   The following ligatures occur: line 1, AM; line 4, AV; line 5, VA.

5 ΤΩΝΟϹΠΡΕϹΒΕΥ
ΤΟΥΑΥΤΟΚΡΑΤΟΡοϺ
ΚΑΙϹΑΡΟϹ [erased]
[erased] ϹΕΒΑϹΤοϒ
ΑΝΤΙϹΤΡΑΤΗΓΟΥΛΥΚϺ
10 ΑϹΚΑΙΠΑΜΦΥΛΙΑϹ
ϽΝΤΩΝ
ΟϹΤΟΥϹΕΑΥΤΟϺ
. . . . .
. . . αν Λουκί[ου θυ-]
γατέρα Π[α]ῦλλα[ν]
γυναῖκα Γαίου Κ[α-]
ριστανίου Φρόν-
5 τωνος πρεσβευ-
τοῦ αὐτοκράτορο[ς]
καίσαρος [Δομ-
ετιανοῦ] Σεβαστοῦ,
ἀντιστρατήγου Λυκ[ί-]
10 ας καὶ Παμφυλίας
. . . . [Φρό]ντων [γυναικὶ]
. . . . . τοὺς ἑαυτο[ύς . . .]

No. 108 is inserted here because it adds to the list of names given in No. 107 that of C. Caristanius Fronto. An inscription of this same C. Caristanius Fronto has been published in the *Bulletin de Correspondance Hellénique*, 1886, p. 46, where he is πρεσβευτὴς αὐτοκράτορος as here, but from our inscription we learn that he was also ἀντιστράτηγος Λυκίας καὶ Παμφυλίας. In lines 7 and 8 of my No. 108 the name of the emperor has been erased. Now the name of Vespasian was never erased, and the remarks of Messrs. Cousin and Diehl (*Bulletin*, as cited p. 47) prove only that the period we have to deal with is that of the Flavian emperors. Vespasian is out of the question for the above reason, and consequently it is probable that the name of Domitian must be restored in No. 108. The *Bulletin*, as cited p. 47, also publishes an inscription of C. Caristanius Paulinus.

No. 109.

Yalowadj.   In wall of the Djami by the market.   Ephemeris
Epigraphica, 1884, *p.* 579, *No.* 1346.   *Copy and impression.*

```
///////////////////////
S E///////////////
A R ⌐ ¯//////////
S A N C T I/////
D O M I N I · N ·
5  A N T O N I N I
A V G D V C E N ᴀ
R I O E T A M V S I ᵒ
S A C · P E R P E T
D E I A E S C ᵥᵢₐ P I
10   P A///////////////
///////////////////////
```

```
·  ·  ·  ·  ·  ·
se . . . . . . .
ar . . . . . . .
sancti[ssimi]
domini n(ostri)
5   Antonini
Aug(usti) ducena-
rio et a musio
sac(ro) perpet(uo)
dei Aescu[la]pi
10   pa . . . . . . .
·  ·  ·  ·  ·  ·
```

Line 1, *EE* omits.
Line 2, *EE* reads AR.
Line 3, *EE* reads SANCI.

Line 8, *EE* reads PE P.
Line 9, *EE* omits VIAPI.

No. 110.

*Yalowadj. In the wall of a school-house. Length, 0.92 m.;
length inside the panel, 0.67 m.; whole width, 0.58 m.;
width inside panel, 0.31 m.* C.I.L. *III.* 289; *Le Bas-
Waddington,* Voyage Archéologique, *III.* 1820. *Copy and
impression.*[1]

```
    C · A R R I O C · F ·
    Q V I R I N A
    C A L P V R N I O
    F R O N T I N O
 5  H O N O R A T O
    C · V · III · V I R · M O N E
    T A L I A A A · E E · Q V A E S
    T O R I C A N D I D' A T O
    P R A E T O R I C A N D I D
10  A V G V R I C O S · P A
    T R O N O C O L · P O S
    T V L · P O P · I N T H E A T R O
    V I C · V E L A B R V S
```

See the minuscule text of No. 112.

No. 111.

*Yalowadj. In the wall of the Djami near the market. Length,
0.92 m.; width, 0.57 m.* C.I.L. *III.* 290; *Le Bas-Waddington,* Voyage Archéologique, *III.* 1819, *from a copy of
Hamilton. Copy and impression.*[2]

---

[1] The reading of lines 7, 11, and 13 is certain. In line 7, A and E, and in line 12, H and E and N and T are in ligature.

[2] The reading of lines 4, 7, 12, 13 is certain. Ligatures are: in line 6, N and E; in line 7, A and E; in line 12, T, H, and E.

```
   C A R R I O C F
   Q V I R I N A
  .C A L P V R N I O
   F R O N T I N O
5  H O N O R A T O
   C · V · III · V I R M O N E
   T A L I A A A ΓΓ Q A E S
   T O R I C A N D I D A T O
   ▨R A E T O R I C A N I
10 ▨▨G V R I C O S P A
   T R O N O C O L · P O S
   T V L P O P · I N T H E A T R O
   V I C · A≃ D I L I C I V S
```

See the minuscule text of No. 112.

No. 112.

Yalowadj. In the foundation of a wall on the side of the Acropolis facing the village of Hissar. It was re-excavated for me by a man who had seen it four years previously, while digging stones for his house. Copy.[1]

```
   C A R R I O C · F
   Q V I R I N A
   C A L P V R N I O
   F R O N T I N O
5  H O N O R A T O
   C V I I I · V I R M O N E
   T A L I A A A ͞I͞I Q A E S
   T O R I C A N D I D A T O
   P R A E T O R I C A N D I D ·
```

---

[1] Ligatures are: in line 7, A and E; in line 12, H and E, T and R.

```
10   A V G V R I C O S · P A
     T R O N O C O L · P O S
     T V L · P O P · I N T H E A T R O
•    V I C · P A T R I C V S
```

C. Arrio, C(aii) f(ilio),

Quirina,

Calpurnio

Frontino

5   Honorato,

c(larissimo) v(iro), triumvir(o) mon-

etali a(uro) a(rgento) a(ere) [f(lando) f(eriundo)], quaes-

tori candid(ato),

praetori candid(ato),

10   auguri, co(n)s(uli), pa-

trono col(oniae), pos-

tul(ante) pop(ulo) in theatro

vic(us) Patric(i)us.

The Consul Suffectus C. Arrius is not mentioned elsewhere ; Waddington thinks he belongs to the third century.

No. 113.

Yalowadj. In the wall of the Djami near the market. Whole length, 1.2 *m.; length inside the mouldings,* 0.80 *m.; whole width,* 0.57 *m.; width inside the mouldings,* 0.37 *m.* C.I.L. *III.* 297 ; *Le Bas - Waddington,* Voyage Archéologique, *III.* 1822 ; *Henzen,* 6156, *from a copy of Hamilton. Copy and impression.*[1]

---

[1] Ligatures are: lines 4, AM; 5, NE; 6, ET, HE; 7, AM, NT; 8, BI; 9, VA; 10, THE; 12, AM, RH; 13, NE, HE; 15, IT; 7, END.

              C N D O T T I O
              D O T T I M A R Y L L I
              N I F I L S E R · P L A N C I
              A N O P A T R · C O L · F L A M ·
        5   Ti V I R Ti Q̄Q̄ · M V N E R · Ti
              E T A G O N O T H E · P E R P ·
              C E R T A M · Q̄Q̄ · T A L A N T ·
              A S I A R C T E M P L · S P L E N D
              C I V I T · E P H E S · E X L I B E
        10 R A L S V A E L E C T · A G O
              N O T H E P E R P A B I M P
              D I V O M A R C O C E R
              T A M · S A C R H A D R I A
              N I O N E P H E S I
        15 P O S T V L P O P V L O
              O B M E R I T · E I V S
            D ᵛ ᴵ ᶜ ᵀ ᵛ ˢ ᶜ ᵛ ˢ D

See the minuscule text of No. 115.

No. 114.

Yalowadj. Ibidem. Length, 1.2 *m.; width,* 0.57 *m.* C.I.L.
III. 296; *Le Bas-Waddington,* Voyage Archéologique,
III. 1822; *Henzen,* Collectio, 6156, *from a copy of Hamil-
ton. Copy and impression.*[1]

              C N D O T T I O
              D O T T I M A R Y L L I
              N I · S E R · P L A N C I A

---

[1] Certain readings are: line 4, ELAM; line 14, POSTVE; line 15, MERT;
line 16, CERMALVS and not GERMALVS. Ligatures are : lines 3, NI; 4, TR;
5, QQ; 6, ET, THE, ER; 7, QQ, NT; 8, TE, ND; 9, IT, HE; 10, VA;
11, THE; 12, RT, AM; 13, NI; 7, PL.

```
      N O P A T R · C O L E L A M
5   ĩĩ V I R · π Q̄Q̄ · M V N E R · ĩĩ
      E T · A G O N O T H E · P E R P · C E R
      T A M · Q̄Q̄ · T A L A N T A S I
      A R C H · T E M P L · S P L E N D
      C I V I T · E P H E S · E X L I B E
10  R Λ L S V A E L E C T A G O
      N O T H E P E R P A B I M P ·
      D I V O M A R C O C E R T A M
      S A C R H A D R I A N I O N
      E P H E S I P O S T V E P O P V L
15  O B M E R T E I V S
      V I C C E R M A L V S ·
```

D   D

See the miniscule text of No. 115.

No. 115.

Yalowadj. The stone is used as a step in the stairway leading to the second story of a house in the Mahallü, called Abudjilar. *It is much worn and almost illegible, and I give it as it looks now. Copy.*

```
      C N D O T T I O
      D O T T I M A R Y L
      I N I F · S E R · P L A N C I
      A N O P A T R · C O L F L A M
5   π V I R ĩĩ Q̓Q M V N E R
      ĩĩ E T A C O N O T H P P R P
      C E R T A M Q Q T A L A N
      A       A R T E M P L S P    N D
      C I V I T E P H E S E X    B E B
10  S V A        T A C O N O T
```

```
P≡R⁊P⁊A S I M P D I V O
M A R C O C E R-T S A C R
H A D R I A N O N E ᴚ H⁊
S I ᴘ O ⁊⁊V⁊⁊O⁊ᴘ⁊V⁊R
15  O R⁊⁊⁊⁊⁊⁊E I V S
     ⁊⁊I C S A L V T A R⁊
```

D   D

Cn. Dottio,
Dotti Marul[l]-
ini fil(io), Ser(gia), Planci-
ano patr(ono) col(oniae), f[l]am(ini),
5 II viro, I[l]q(uin)q(uennali), muner(ario)
II [e]t [ag]onoth(etae) p[e]rp(etuo)
certam(inis) q(uin)q(uennalis) talan(tiaei),
A[s]i[a]r(chae) temp[l](orum) sp[le]nd(idissimae)
civit(atis) Ephes(inae) ex [l]ibe[r](alitate)
10 sua, [ele]ct(o) a[g]onot(hetae)
p[e]rp(etuo) a[b] imp(eratore) divo
Marco cert(aminis) sacr(i)
Hadrian[i]on E[p]he-
si [post]u[l](ante) populo
15 ob [merit(a)] eius
[v]ic(us) Salutar[is].
        D(ecreto) D(ecurionum).

The incompleteness of Hamilton's copies of the two Dottius inscriptions (Nos. 113 and 114) is to be ascribed to his failure to
notice the ligatures.

Concerning the ἀγῶνες ταλαντιαῖοι πενταετηρικοί, see Le Bas-
Waddington, *Voyage Archéologique*, 1209, and *C.I.G.* 3208.

Concerning the Asiarchs at Ephesus, see Le Bas-Waddington,
Voyage Archéologique, 158 a, 885, *C.I.G.* 2965, 2987 b; *Ephemeris
Epigraphica*, I. p. 200–214; *Monatsberichte der konigl. preussischen
Akademie der Wissenschaften*, 1874, p. 12; *Revue Archéologique*,

1874, XXVIII. p. 10; *Bulletin de Correspondance Hellénique*, 1880, p. 375; 1878, p. 595; 1883, p. 264 and p. 450; 1886, p. 151; *Academy* [London], Aug. 11, 1883; *Papers of the American School at Athens*, Vol. I. p. 103; but *above all*, see the exhaustive study of Lightfoot, *Apostolic Fathers*, Part II. Volume II. Section II. pp. 987–998.

Concerning the Ἀδριάνεια, see *C.I.G.* 2987 *b*, 3208.

No. 130 above [= *C.I.L.* III. 296; Le Bas-Waddington, *Voyage Archéologique*, 1822] certainly has *Cermalus* as the name of the *vicus*, and consequently M. Waddington's conjecture of *Germanus* is untenable. Professor Mommsen tells me that Κέρμαλος is the writing of the Greeks and that Cermalus is the only true form, though not acknowledged by modern scholars. The known *vici* of Antiochia now number six, the names of which are given in the last six inscriptions, viz. *Tuscus, Cermalus, Aedilicius, Velabrus, Patricius, Salutaris*. It is a singular fact that the modern city of Yalowadj is composed of twelve[1] *vici* — called *Mahallülar* [Mahallü being the Arabic word for "*Quarter*"] — and these modern *vici* may be an inheritance from antiquity.

No. 116.

Gemen [*Yemen*], *about one hour to the southeastward of Yalowadj. In the wall of the Djami. Length,* 1.22 *m.; width,* 0.62 *m. Copy and impression.*

```
C · F · S E R
S A T V R N I N O
P R A E F · F A B R · Q ·
Π V I R · V N I V E R S O ·
P O S T V L A N T E · P O P V L O
O B · A E Q V A M · E T · I N T E
G R A M · I V R I S · D I C T I O
N E M
```

---

[1] I gave the erroneous number of *five* in my *Preliminary Report*, p. 11.

. . . .

C(aii) f(ilio), Ser(gia),
Saturnino,
praef(ecto) fabr(um), q(uaestori)
(duum)vir(o) universo
postulante populo
ob aequam et inte-
gram iuris dictio-
nem.

No. 117.

Yalowadj.   *Fragment in the wall of the Djami by the market. Copy and impression.*

. . . .

[p]rovinci[ae Syri-]
  [ae] Coele[s . . . .]
  [provi]nciae As[iae]
  . . . .]et patron[o coloniae]
  [o]b m(erita) e(ius) pos[tu-]
  [lante p]opulo . . . .

No. 118.

Hissar. In the wall of a Djami. Copy.

```
P R O C V L C▒▒▒▒
A L A · A V G G E▒▒
M A N I C A
H ·        C ·
```

```
.  .  .  .  .  .  .
Procul[o . . . . . . .
ala Aug(usta) Ge[r]-
manica
h(onoris) c(ausa).
```

No. 119.

Yalowadj. In the western cemetery. Copy.

The stone has been cut circularly as if for a round building; see the remarks after No. 84. It is 1.18 m. long; width at one end, 0.55 m.; at the other end it is 0.15 m. wide. Under the arc of the circle are the words

```
V I V I
V I V I S
```

in large letters, and nothing else.

No. 120.

Yalowadj. Column serving as one of the four supports to the roof of the Medressi near the military prayer enclosure. C.I.L. *III.* 303 ; *Le Bas-Waddington*, Voyage Archéologique, 1824 ; Ephemeris Epigraphica, 1884, *p.* 575, *No.* 1342. *Copy.*

```
V · V · P E T I L I A · M · F
T E R T I A · S I B I · E T
M · P E T I L I O · P A T R I ·
Z Ⲱ C A Π Є T I Λ I A T Є P T I A
Є A Y T H K A I M A P K Ⲱ
Π Є T I Λ I Ⲱ Π A T P I
```

V(iva) v(ivis). Petilia, M(arci) f(ilia),
Tertia sibi et
M(arco) Petilio patri.
Ζῶσα Πετιλία Τερτία
ἑαυτῇ καὶ Μάρκῳ
Πετιλίῳ πατρί.

Line 1, *EE* omits the points.   Line 5, *EE* reads ⬚AYTIA.
Line 4, *EE* ΠΕΠΔIA.   Line 6, *EE* omits ΠΕ.

It will be noticed that this bilingual inscription renders the Latin VV by Ζῶσα. Accordingly VV must stand for *viva vivis*.

For a fourth inscription of Antiochia Pisidiae, commencing with VV, see *Papers of the American School of Classical Studies at Athens*, Vol. III. No. 358.

No. 121.

Yalowadj. In the wall of the Medressi near the soldiers' prayer enclosure. Ephemeris Epigraphica, 1884, *p.* 580, *No.* 1353. *Copy.*[1]

```
V·V·RVBPIA        TEPT▨▨▨
. FR∧TRI   [gable]   E ▨▨▨▨
```

No. 122.

Yalowadj. Quadrangular cippus in the cemetery of Abudjilar. Length, 1.45 *m.; width,* 0.51 *m. Copy and impression.*

```
    PIETATI
  AVGVSTORVM
  NOSTRORVM
 VALDIOGENESVP
▨▨▨IESPROVINPISID
```

---

[1] In line 2, TR are in ligature.

Pietati
Augustorum
nostrorum
V[a]l(erius) Diogenes v(ir) p(erfectissimus)
[praes]es provin(ciae) Pisid(iae).

No. 123.

Hissar, a village half an hour east of Yalowadj. Epistyle block ornamented with the eggstaff; lower facet, 0.10 m.; middle facet, 0.125 m.; top facet, with the eggstaff, 0.175 m. Height of letters on middle facet, 0.11 m.; on bottom facet, 0.09 m. The block now stands endwise as a doorpost, and the commencement of the inscription cannot be gotten as the stone is buried. No Alpha bars. Copy.

ROPITIAMAIESTATEDDNN · SE
FVNDAMENTO · DIOGENES · V · P

. . . . [p]ropitia maiestate d(ominorum) n(ostrorum) Se . . .
. . . . [a] fundamento Diogenes v(ir) p(erfectissimus),
[praeses provinciae Pisidiae].

No. 124.

Hissar. Fragment in the street. Copy.

D N
O N S T A N
ICTO
GEN

D(omino) n(ostro) [Imp(eratori) Caes(ari) C-]
onstan[tino P(io) F(elici) inv-]
icto [Aug(usto) . . . . . .
Dio]gen . . . . . .

No. 125.

Yalowadj. In the wall of a house. Length, as far as visible, 0.92 *m.; width,* 0.72 *m.; height of letters,* 0.08 *m. Copy.*

C N · P O M P E I C▨
C O L L E G A E
P A T R O N O C O▨

D   D

Cn. Pompei[o]
Collegae
patrono co[l(oniae)]
d(icreto) d(ecurionum).

Cn. Pompeius Collega was legatus Galatiae under Vespasian, see *C.I.L.* III. 306, and Le Bas-Waddington, *Voyage Archéologique,* 1814 *b.*

No. 126.

Yalowadj. In the wall of a Kouak. Copy.

N O N I V S O P T A T V S▨
N O N I A E P A V L I N A · F S▨

Nonius Optatus
Nonia[e] Paulina[e], f(iliae) s(uae).

No. 127.

Yalowadj. Immense block serving as a step in a stairway. C.I.L. *III.* 302; *Le Bas-Waddington,* Voyage Archéologique, *III.* 1191, *from a copy of Arundell. Copy.*

T I · C L A V D I O
P A V L L I N O
P H I L O S O
P H O · H E R O

Line 2.   PAVLINO is the reading of the publications referred to.

No. 128.

Yalowadj. In the western cemetery. Length, 1.12 m.; width, 0.50 m. Broken at the left; top, bottom, and right side whole. Ephemeris Epigraphica, 1884, *p.* 579, *No.* 1347. *Copy.*

```
       CIATRVNO
       ORNVTIFILPA
       OCOLQVIEXLIIIF
```

No. 129.

Yalowadj. In the wall of the soldiers' prayer enclosure. I have a note that the fifth and sixth letters in line 1 may be LL. Ephemeris Epigraphica, 1884, *p.* 580, *No.* 1349. *Copy.*

```
       OSEXILF
       NO
       IVSAMICO
       KCAVSA
```

No. 130.

Yalowadj. In the western cemetery. Quadrangular stone with moulding. Copy.

M·CORNELIVS·M·F·

No. 131.

Hissar. In the wall of a house. Copy.

```
SEX·A·PPVLE
CASSANDRI
```

No. 132.

Yalowadj.   In the court of a Kouak.   Copy.

[uncut] C I [uncut]
P A V L L I N A
S A C▨▨▨▨▨

No. 133.

Yalowadj.   Fragment in the cemetery of Abudjilar.   Copy.

▨▨▨▨▨▨
▨P O N I▨▨▨
▨▨E X · T E S T▨
▨▨▨▨S V B▨
▨▨▨▨▨▨

No. 134.

*Yalowadj.   Fragment of an epistyle in the western
cemetery.   Copy.*

▨A S A C▨

No. 135.

*Yalowadj.   In the wall of the Medressi near the prayer
enclosure.   The stone is broken on all sides, but still little
seems to be gone.   Copy and impression.*[1]

---

[1] Ligatures occur: line 7, MH; line 9, NN.

```
KOΛONEIΛ Y
KAIΓ·ΦΛAIOYBA
BIANOYIΠΠO
PϾMAIϾNAPXI
PEϾNΔIABIOY
TOYΠATPIOY
ΘEOYMHNO
CIOYΛIONΔO
NONNOI⋯⋯
```

. . . .

' . . .[τῆς] Κολονεί[ας ν . . . .

. . . . . . καὶ Γ. Φλα(βίου) 'Ιου(λίου) βα[ιβίου?]

. . .[Φλα]βιανοῦ? ἱππό[του . . . .

. . .[τῶν] 'Ρωμαίων ἀρχι-

[ιε]ρέων διὰ βίου

. . . . . . τοῦ πατρίου . . .

. . . . . . θεοῦ Μηνὸ[ς . . .

. . . . . . ς 'Ιούλιον Δό[μ?]

νον Νο[ννον?]

. . . .

No. 136.

*Yalowadj. Horned altar in the cemetery of Abudjilar.
Copy and impression.*[1]

---

[1] Ligatures occur: line 2, NT; line 4, MN; line 10, ON; line 13, TH; line 15, ΠP.

Side *A.*

KEIΠIOC
CYNTPOΦOCKAI
EYTYXIAEYTYXI
ANWTEKNWMNEI
5 ACXAPIN

Side *B.*

WCΦYTONAPTIΘA
ΛECΔPOCEPOICΠA
PANAMACINAY'ZON
WCPOΔONAPTIΦY
10 ECΠPOΦANENKAΛON
ANΘOCEPWTWN
OYTWCΔHKAIΠAI
ΔAXYTHKATAΓAIAKA
ΛYΠTEIZWTIKONO
15 ΓΔOONHΛIKIHCΠPO
ΛABONTENIAYTON

Side *A.*

Κείπιος
Σύντροφος καὶ
Εὐτυχία Εὐτυχι-
ανῷ τέκνῳ μνεί-
ας χάριν.

Side *B* is composed of four hexameters, thus:

ὡς φυτὸν ἀρτιθαλὲς δροσεροῖς παρὰ νάμασιν αὐ[ξ]ον,
ὡς ῥόδον ἀρτιφυὲς προφανὲν καλὸν ἄνθος ἐρώτων,
οὕτως δὴ καὶ παῖδα χυτὴ κατὰ γαῖα καλύπτει
Ζωτικὸν ὄγδοον ἡλικίης προλαβόντ' ἐνιαυτόν.

No. 137.

Yalowadj. Panel on a rough stone by the side of a water conduit in the street. Length, 1.08 m.; width, 0.90 m. Copy.

```
  Λ · Μ Α Λ Ι Ο C Φ Λ Α Κ Ο C
  Κ Α Ι Γ · Μ Α Λ Ι Ο C Μ Α Ξ Ι Μ Ο C
  Α Δ Ε Λ Φ Ο Ι Ε Α Υ Τ Ο Ι C Κ Α Ι
  Τ Ο Ι C Ι Δ Ι Ο Ι C Κ Α Ι Λ · Μ Α Λ Ι Ω
5 Μ Α Ξ Ι Μ Ω Ν Ο Μ Ι Κ Ω Τ Ε Κ Ν Ω
  Γ Λ Υ Κ Υ Τ Α Τ Ω Κ Α Ι Τ Ο Ι C Ι Δ Ι Ο Ι C
  Γ Ο Ν Ε Ι C Ι Μ Ν Η Μ Η C Χ Α Ρ Ι Ν
```

Λ(ούκιος) Μάλιος Φλάκος
καὶ Γ(άϊος) Μάλιος Μάξιμος
ἀδελφοὶ ἑαυτοῖς καὶ
τοῖς ἰδίοις καὶ Λ(ουκίῳ) Μαλίῳ
5 Μαξίμῳ νομικῷ τέκνῳ
γλυκυτάτῳ καὶ τοῖς ἰδίοις
γονεῖσι μνήμης χάριν.

No. 138.

Yalowadj. Stele in the court of a Kouak. Copy.[1]

```
  Α Υ Ρ Η Λ Ι Α Ο Υ Α Λ Ε Ν Τ Ι Λ Λ Α
  Α Υ Ρ Η Λ Ι Ω Μ Α Κ Ε Δ Ο Ν Ι
  Α Ν Δ Ρ Ι Γ Λ Υ Κ Υ Τ Α Τ Ω Τ Η Ν C Τ Η Λ Η Ν
  Α Ν Ε C Τ Η C Α Μ Ν Η Μ Η C Χ Α Ρ Ι Ν Ο C
5 Δ Ε Α Ν Ε Π Ι Β Ο Υ Λ Ε Υ C Ι Τ Η Ν C Τ Η Λ Η Ν
  Ε C Τ Α Ι Α Υ Τ Ω Π Ρ Ο C Τ Ο Μ Ε Γ Ε
  Θ Ο C Τ Ο Υ Θ Ε Ο Υ
```

---

[1] Ligatures occur: lines 3, ΤΗΝ, ΤΗ, ΗΝ; 4, ΤΗ, ΜΝΗΜΗ; 5, ΤΗΝ, ΤΗ, ΗΝ.

Αὐρηλία Οὐαλέντιλλα
Αὐρηλίῳ Μακεδόνι
ἀνδρὶ γλυκυτάτῳ τὴν στήλην
ἀνέστησα μνήμης χάριν· ὃς
5 δὲ ἂν ἐπιβουλεύσι τὴν στήλην·
ἔσται αὐτῷ πρὸς τὸ μέγε-
θος τοῦ θεοῦ.

Concerning curses invoked on violators of tombs, see *Papers of the American School of Classical Studies at Athens*, I. p. 84.

The name Οὐαλέντιλλα occurs in an inscription of Iconium, *C.I.G.* 3996, and in an inscription of Kirili Kassaba, our No. 189.

No. 139.

Yalowadj.   Quadrangular cippus: height, 1.10 m.; width, 0.51 m.; height inside the mouldings, 0.53 m.   Le Bas-Waddington, Voyage Archéologique, 1189, *from a copy of Falkener first published by Henzen in the* Annali dell' Instituto.   *Copy.*[1]

```
   Λ·ΚΑΛΠΟΥΡΝΙΟΝ
   ΡΗΓΕΙΝΙΑΝΟΝ
   ΤΟΝΛΑΜΠΡΟΤΑΤΟΝΣΥ▦
   ΚΛΗΤΙΚΟΝΥΙΟΝΚΑΛ
5  ΠΟΥΡΝΙΟΥΡΗΓΙΝΙΑΝΟΥΤΟΥ
   ΛΑΝΠΡΟΤΑΤΟΥΥΠΑΤΙΚΟΥ
   ΟΥΑΠΙΟΣΤΑΤΙΑΝΟΣΜΑΡΚΕΛΟ   C
   ΔΥΑΝΔΡΙΚΟΣΑΡΧΙΕΡΕΥΣΔΙΑ
9  ΒΙΟΥΤΟΥΕΠΙΦΑΝΕΣΤΑΤΟΥΘΕΟΥΔΙΟΝΥΣΩ▦
```

---

[1] In line 3 ΜΠ are in ligature.

Λ(ούκιον) Καλπούρνιον
Ῥηγεινιανὸν
τὸν λαμπρότατον συ[ν]-
κλητικὸν υἱὸν Καλ-
5 ᵘπουρνίου Ῥηγινιανοῦ τοῦ
λανπροτάτου ὑπατικοῦ,
Οὔλπιος Τατιανὸς Μάρκελος
δυανδρικὸς, ἀρχιερεὺς διὰ
0 βίου τοῦ ἐπιφανεστάτου θεοῦ Διονύσ[ου].

No. 140.

Yalowadj. On a sarcophagus in the court of the Djami nearest the barracks. Part of the inscription (lines 1–4) is given in C.I.G. 3981. Copy.

```
ΚΑΤΑΣΥΝ⚡ΩΡΗΣΙΝ
ΤΟΥΑΞΙΟΛΟΓΩΤΑΤΟΥ
ΑΥΡΗΛΙΟΥΚΑΝΔΙΔΟΥ
ΗΣΟΡΟΣΕΤΕΘΗΕΝΗ

ΚΑΤΕΤΕΘΗΣΩΜΑ
ΤΑΑΥΡΗΛΙΑΣΣΤΡΑ
ΤΟΝΕΙΚΗΣ
ΚΑΙΟΡΤΗΣΙΑΝΟΥ
ΜΑΞΙΜΟΥΑΝ
ΔΡΟΣΑΥΤΗΣ
```

Κατὰ συν[χ]ώρησιν
τοῦ ἀξιολογωτάτου
Αὐρηλίου Κανδίδου
ἡ σορὸς ἐτέθη, ἐν ᾗ

κατετέθη σώμα-
τα Αὐρηλίας Στρα-
τονείκης
καὶ Ὀρτησιανοῦ
Μαξίμου ἀν-
δρὸς αὐτῆς.

No. 141.

Yalowadj. Fragment in the wall of the Djami of Abud-jilar. Letters faint and blurred. Copy and impression.

Ω Α Ν Τ Ι Σ Γ Ε Ι Ν
Χ Α Ι Ρ Ε Φ Α Ν Η Σ Μ Ο Ι
Γ Ω Ν Τ Ο Υ Τ Ο Τ Ρ Ο
Π Α Ι Ο Ν Ο Ρ Α Σ
Μ Ε Λ Υ Κ Α Ο Ν Ι Η
Θ Α Ν Α Τ Ο Ν Μ Ι Τ Ο Ι
Η Δ Ε Κ Α Λ Υ Ψ Α Ν
Ξ Ε Ι Ν Ο Ν Κ Α Ι Π Ο
Λ Ι Ο Σ Κ Α Ι Τ Ο Π Ο Υ
Ω Κ Ε Χ Υ Μ Α Ι
Ο Υ Μ Ο Ι Κ Α Ι Τ Ο Δ Ε
Σ Η Μ Α Ε Τ Ε Ι Δ Ε Κ Α Τ Ω
Μ Ε Τ Ε Π Ε Ι Τ Α

Χαῖρε, Φάνης, μοι[ρ]ῶν τοῦτο τροπαῖον ὁρᾷς
[Ὢν] με Λυκαονίη θάνατον μίτοι ἠδὲ κάλυψαν
[Ξ]εῖνον καὶ πόλιος καὶ τόπου ᾧ κέχυμαι
Οὗ μοι καὶ τόδε σῆμα ἔτει δεκάτῳ μετέπειτα.

No. 142.

Yalowadj. Stele with gable in a tanyard near the cemetery of Abudjilar. Copy.

```
ΑΡΤΕΙΜΕΙϹΙΑΜΑ
ΝΤΟΥΝΤΗϹΥΝ
ΒΙШΜΝΕΙΑϹΧΑ
ΡΙΝΤΟΝΘΕϹ
ΙΝϹΥΜΗΛΔΙ
ΚΗϹΕΙϹ
```

'Αρτεμεισία Μα-
ντοῦν τῇ συ[ν]-
βίῳ μνείας χά-
ριν· τὸν θέσ-
ιν σὺ μὴ [ἀ]δι-
κήσεις.

Note the gender of the article in lines 2 and 4.

No. 143.

Yalowadj. Stele with gable so high up in the wall of a house that I could see it only with difficulty. Copy.

```
ΑΘΑΝΑΤΟΥΨΥΧΗϹ
ϹΤΗΛΗΝΑΝΕΘΗ⸖
ΚΑΙΟΥΚΟΥΝΔΟϹ
ΤΥΝϹΒШΑΛΕΞΑΝ
ΔΡШΜΝΗΜΟϹΥΝΗϹ
ΕΝΕΚΕΝ
```

'Αθανάτου ψυχῆς
στήλην ἀνέθη
Κ[λ](αύδιος) 'Ιούκουνδος
τ[ῷ? συμβίῳ?] 'Αλεξάν-
δρῳ μνημοσύνης
ἕνεκεν.

In lines 4–5 we naturally expect the name of a woman, not that of a man.

No. 144.

Yalowadj. On a sarcophagus in the yard of a house. Copy.

On the top mculding of the side in one long line is the following: —

A.

ΕΙΔΕΤΙΕΤΟΝΤΟΠΟΝΗΛΑΡΝΑΚΑΤΗΝΔΕΑΔΙΚΗΣΕΙΟΡΦΑΝΑ///////ΕΡΗΜΟΝΕΝΤΥΡΙΠΤΥ//////
ΤΕΚΝΑΛΙΠΟΙ////////ΙΡΑΓΩΟΙΤΟ

Εἰ δέ τις τὸν τόπον ἢ λάρνακα τήνδε ἀδικήσει,
ὀρφανὰ τέκνα λίποι, [χῆρον βίον, οἴκου] ἔρημον,
ἐν πυρὶ π[ά]ντα δάμοιτο, κακῶν ὑπὸ χείρ[ας ὄλ]οι[τ]ο.

In a panel occupying the centre of the side of the sarcophagus is this : —

B.

```
CΛΛINACATOYPNIN░
ΓYNHAYTOYZWNTEC
KAIΦPONOYNTECEAY
TOICTHNCOPONKATE
CKEYACANMNH
MHCXAPIN░
```
5

Σαλῖνα Σατουρνῖν[α ἡ]
γυνὴ αὐτοῦ ζῶντες
καὶ φρονοῦντες ἑαυ-
τοῖς τὴν σορὸν κατε-
σκεύασαν μνή-
μης χάριν.

Apparently the inditer of the inscription was ignorant of the fact that the words between ὀρφανὰ τέκνα and χεῖρας ὄλοιτο form two hexameters, else he would have thrown the first part into verse (see *C.I.G.* 4000, also 3862, 3875, 3990 *k*). The name of the husband is wanting in the panel; an oversight of the stonecutter.

No. 145.

Yalowadj.   Small horned stele in the court of a house.   Copy.

```
OYIPIAΔOMNA
ZϾCIMϾANΔPIΓΛY
KYTATϾMNHMHC
XAPIN
```

Οὐιρία Δόμνα
Ζωσίμῳ ἀνδρὶ γλυ
κυτάτῳ μνήμης
χάριν.

The name Οὐιρία is new and is probably indigenous.

No. 146.

Yalowadj.   Stele surmounted by a gable.   In the western cemetery.   Copy.

ΠΕΙϹШΝΤΙ
ΤШΤШΙΔΙШ
ΑΔΕΛΦШΙΕ
░ΠΟΛΕΙΤΗ
░ΜΗϹ░

Πείσων Τί-
τῳ τῷ ἰδίῳ
ἀδελφῷ Ἱε-
[ρα]πολείτῃ
[μνή]μης [χάρω].

It is, of course, impossible to determine which city is meant as the native town of Titus; if the city in the Sandükli Ovasü be the one meant, then Ἱε[ρο]πολείτῃ must be restored (see *Journal of Hellenic Studies*, 1882, p. 340 sqq., *Bulletin de Correspondance Hellénique*, 1882, p. 519); if the city in Phrygia Pacatiana be meant, then Ἱε[ρα]πολείτῃ would be correct.

No. 147.

Yalowadj.   Large stone [1.49 m. × 0.80 m.; height of letters, 0.8 m.] serving as an abutment for the wooden staircase of a house. The feet of the staircase rest on the stone as indicated below.   Copy.

░ΟΙΝΤΟϹΜΟΥΝΗ
ΟϹΕΥΤΥΧΗϹ
ΚΟΙΝΤΟΥΜΟΥΝΗ
ΤΙΟΥΠШΛΙΩΝΟϹ
ΠΡΑΓΜΑΤΕΥΤΗϹ

[Κ]όϊντος Μο[υν]ή-
[τι]ος Εὐτυχ[ὴς]
Κοΐντου Μουνη-
τίου Πωλίωνος
πραγματευτής.

Q. Munatius Pollio is the Latin form of the name. The family is mentioned in an inscription of Yalowadj in *Papers of American School at Athens*, Vol. III. No. 352.

No. 148.

Yalowadj. Broken panel in the court of a Konak. Copy.[1]

▨▨▨▨▨\ΕΟΕΕΡΓΟΝΕΗΙ
▨▨▨▨▨ΛΥΒοΥΛШΙ [uncut]
▨▨▨▨ιΟΠΡΟΤΕΡШΝΗΝΥΕΕ
▨▨▨▨ΕΕΜΟΠΟΛШΝ [uncut]
▨▨▨ιΡΕΕΘΡΟΝΑΝΑΝΤΑΕοιοΕ
▨▨▨ΡШΝ+ΕΕΧШΡΟΥΕΑΓΑΓΕΙΝ
▨▨▨ΥΑΛΕΟΥΕΠΡΟΤΕΡΟΝ

No. 149.

Yalowadj. Slab, length, 1.8 m.; width, 0.71 m. In the court of a house. Right side defaced. Copy.

†ΟΡΑΙΕΤΟΔΕΡΓΟΝΗΜΙΙΙΟΝ▨▨▨▨
Π⏾ΕΔΑΥΙΛΗ
ΝΥΜΦ⏾ΝΟΡΗΓΕΙΤΙΙΠΟΛΕ▨▨▨▨
ΤΑΝΑΝΑΤΑ
ΕΟΦΟΕΔΙΚΛΕ▨▨▨▨▨▨▨▨▨▨
ΕΤΙΑΤΙ▨▨▨
ΗΓΕΡΕΝΑΥΤΟΕΙΕΙ▨▨▨▨▨▨▨▨
ΙΝΔ▨▨▨▨

[1] I have a marginal note that the last P in the last line looks somewhat like a φ.

No. 150.

Yalowadj. Quadrangular cippus in the western cemetery.
Copy.

```
N E O Y I O I
C A T O Y P N E I N O C
⌐ Λ I O P E B I A I Φ H
Ɔ O I . ω X P Y C O v A C
5   Λ ω Γ Λ Y Ϙ Y T A T ω
Θ P E Π T ω M N H M H Ɔ
X A P I N
```

. . . . .

Σατουρνεῖνος

. . . . .

. . . . .

5   λῳ γλυκυτάτῳ
θρεπτῷ μνήμης
χάριν.

No. 151.

Yalowadj. In the wall of a Konak. Copy.

```
▨▨▨▨▨▨▨▨▨
Δ I A C N ω ı T I ω
I Δ I ω C Y N B I ω
Μ N H Μ H C X A
P I N
```

. . . . . . [Εὐσ-]
δία ['E]νω[τ]τίῳ?
ἰδίῳ συνβίῳ
μνήμης χά-
ριν.

No. 152.

*Yalowadj. In the court of a house by the fountain of
Abudjilar. Copy.*

```
IOCZWTIKOC
IOYΛIAMATPW
NHΓΛYKYTA
TH
```

. . . . . [ʾΙούλ-?]

ιος Ζωτικὸς

ʾΙουλίᾳ Ματρώ-

νῃ γλυκυτά-

τῃ.

No. 153.

Yalowadj. In the wall of a house. Copy.

```
       ΔΙΟ
      ΚΑΙ
   OYNΔAMENOI
TOYEAYTOIΣKAI
TOIΣEΓΓONOIΣ
```

. . . .

. . . . Δ]ιό[δωρος τοῦ]

[δεῖνος] καὶ [ʾΙουλία?]

[Σεκ]οῦνδα Μενοί-

του ἑαυτοῖς καὶ

τοῖς ἐγγόνοις.

No. 154.

Yalowadj.   In the wall of a house.   Copy.

```
////////////////////////////////
////O C T E K N ш
////N H M H C X A
  ⅏ P I N ⅏
```

```
    ·   ·    ·   ·   ·
-ος τέκνω
μνήμης χά-
ριν.
```

For additional inscriptions of Yalowadj (Antiochia Pisidiae) and the region of country around it, see the *Papers of the American School at Athens,* Vol. III. pp. 218–278.

The ruins of Antiochia Pisidiae have been sufficiently described by former travellers, so that further mention by me is superfluous.

June 20.   Yalowadj to Ak Shehir, 5 h. 42 m.   Leaving Yalowadj the road leads over the Acropolis of Antiochia in 30 m. to Hissar. Thence we go up a narrow gorge, down which comes a stream of water.   In 43 m. from Hissar the gorge divides: we follow neither arm, but ascend in abrupt zigzags the steep and rugged spur of the mountain which lies between the two gorges.   A new road was just being constructed across the Sultan Dagh between Yalowadj and Ak Shehir.   It was finished in spots, and hundreds of men were still at work on it.   It is of course superior to the old road, but the new road will still remain a trying and laborious one.   This road between Yalowadj and Ak Shehir has hitherto been thought to be a pass. But it is a pass only in so far as deep gorges lead up to the great backbone of Sultan Dagh on either side (see the large map in Vol. III. of *Papers of the American School at Athens*).   The mountain sends off ridges without number at right angles to the mountain chain, and any two opposite gorges may be called a pass with as much propriety as the two which lie on opposite sides of the mountain between Yalowadj and Ak Shehir.   The road reaches the great plain of Philomelium one hour northwest of Ak Shehir, and consequently it does not go down the gorge, at the mouth of which the city of Ak Shehir lies.

No. 155.

Ak Shehir (Philomelium). Diminutive cippus now, in the possession of Dr. Diamantides in Konia. Copy.

N A O C

O K H N O C

Δ Ε Ι Ζ Ε Μ Ε Τ Α Ε Ι Ν

Ε Υ Χ Η Ν

But few remains of Greek antiquity are to be found at Philomelium; but, on the other hand, the traveller is surprised by some Seldjuk ruins of exquisite beauty. The accurate workmanship displayed, even in the execution of details, will compare favorably with Greek buildings of a good period.

At Ak Shehir I was joined, as had been previously arranged, by my friend, Professor J. H. Haynes, then of Robert College, Constantinople, now of the Central Turkey College, Aintab, Syria. Mr. Haynes accompanied me as photographer during the rest of the journey. My travelling-outfit had been left at Smyrna, and I had not fared well thus far. The advent of Mr. Haynes and the outfit was hailed with delight; for henceforward we could have substantial food, on which depends in great measure the success of an expedition like this.

June 21. Ak Shehir to Engilli, 1 h. 24 m. From Ak Shehir my route lay along the foot of Sultan Dagh in a southeasterly direction to Daghan Hissar. This region is very populous, and what is a blank mountainous space on the old maps is in reality a plain full of prosperous villages (see the large map in Vol. III. of the *Papers of the American School at Athens*).

June 23. Engilli to Daghan Hissar, 5 h. 5 m. My route lay along the foot of Sultan Dagh, and is indicated by the red line on the map. I found no inscriptions in the villages between Engilli and Kara Agha, but the topographical results were abundant.

No. 156.

Kara Agha, a village one hour northwest from Daghan Hissar. Quadrangular cippus in the wall of the Djami. Copy and impression.

```
Α Υ Ρ Π Α Τ Ρ Ο
Κ Λ Η Ϲ Κ Α Ι Δ Ο Υ
Δ Α Ϲ Ο Υ Ϲ Ο Υ Υ
Ι Ω Ο Ρ Ο Φ Υ Λ Α
Κ Ι Ι Ϲ Φ Α Γ Ε Ν Τ Ι
Υ Π Ο Λ Η Ϲ Τ Ω Ν
Μ Ν Η Μ Η Ϲ Χ Α Ρ Ι▨
```

Αὐρ. Πατρο-
κλῆς καὶ Δου-
δᾶς Οὔσου υ-
ἱῷ ὀροφύλα-
κι⟨ι⟩ σφαγέντι
ὑπὸ λῃστῶν
μνήμης χάρι[ν].

The name of the son, Οὔσου, must be indeclinable. For ὀροφύλακι, see No. 65.

No. 157.

Kara Agha. Quadrangular cippus in the wall of the Djami. Copy and impression.

```
▨▨▨▨▨▨▨▨▨▨
Τ Ε Κ Ν Ο Μ Ν Η Μ Η
▨Χ Α Ρ Ι Ν Κ Α Ι Ε Α Υ Τ Ο
▨Ο Ν Α Ν Ε Ϲ Τ Η Ϲ Ε
```

['Ο δεῖνα τῷ δεῖνα]
τέκνῳ μνήμη-
[s] χάρω καὶ ἑαυτῷ
[ἐκ τῶν ἰδί]ων ἀνέστησε.

No. 158.

Kara Agha. Phrygian door in the wall of the Djami.
Copy and impression.

ΑΥΡΗΛΙΟΣΜΝΗΣΙΘΕΟΣΥΕΙΟΣ
ΠΑΠΑΔΟΣ·)ʿΟΥΡΜΑΑΝΕΣΤΗ▨
ΕΝΤΗΗΔΙΑΓΥΝΕΚΙΒΑΣΙΑΙΣΗ▨
▨ΝΗΜΗ▨ΧΑΡΙΝ

Αὐρήλιος Μνησίθεος νεὶὸς
Πάπαδος [Κ]ουρμᾶ ἀνέστη[σ]-
εν τῇ ἠδίᾳ γυνεκὶ Βασῖ? Αἴσῃ
[μν]ήμη[ς] χάρυ.

Βασὶς is a new name, so far as I can find out.
Note ἠδίᾳ for ἰδίᾳ.

No. 159.

Kara Agha. Quadrangular cippus in the wall of the
Djami. Copy and impression.

ΑΥΡΑ – ΣΚΑ⌣▨
ΔΙΟΥΤΗΓΛΥΚΥΤΑΤ
ΗΜΟΥΣΥΜΒΙШΑΥΡ
ΚΥΡΙΑΜΕΤΑΤΟΥΑ
ΝΕΨΙΟΥΑΥΤΟΥΑ
ΥΡΜΑΡΚΟΥΑΝΕΣ
ΤΗΣΑΝΜΝΗΜΗ
ΣΧΑΡΙΝ [uncut] ΤΙΣ
ΑΝΠΡΟΣΟΙΣΕΙΧΕ
ΡΑΤΗΝΒΑΡΥΦΘΟΝ
ΟΝΚΕΙΝΟΣΔΕΑ
ШΡΟΙΣΠΕΡΙΠΕΣ
ΟΙΤΟΣΥΜΦΟΡΑΙ
Σ

Αὐρ. Ἀ[β]άσκα[ντος?]
Δίου τῇ γλυκυτάτ-
ῃ μου συμβίῳ Αὐρηλίᾳ
Κυρίᾳ μετὰ τοῦ ἀ-
νεψιοῦ αὐτοῦ Α-
υρ. Μάρκου ἀνέσ-
τησαν μνήμη-
ς χάριν.
τὶς ἂν προσοίσει χε[ῖ]ρα τὴν βαρύφθονον
κεῖνος δὲ ἀώροις περιπέσοιτο συμφοραῖς.

The inscription is closed by two iambic trimeters. Cf. *Mittheil-ungen d. Deutsh. Arch. Inst. in Athen*, 1885, p. 17.

No. 160.

Kara Agha. Ornamented Stele in the wall of the Djami. See my Preliminary Report, *p. 10. Copy and impression.*

```
A Y P H Λ Є I O C Z Ѡ
T I K O C Π A Y Λ Є I N O Y
A Δ▨▨N O Π O Λ Є I T H
C T ▨C Y N B I Ѡ A Y
P H▨▨Λ Δ A Γ Λ Y K Y T A
T H Μ N H Μ H C X A P I N
```

Αὐρήλειος Ζω-
τικὸς Παυλείνου
Ἀδ[ρια]νοπολείτη-
ς τ[ῇ] συνβίῳ Αὐ-
ρη[λίᾳ] Δᾷ γλυκυτά-
τῃ μνήμης χάριν.

For the name Δᾶ, see No. 168.

Paulinus is mentioned as a citizen of Hadrianopolis: this inscription does not locate the city, but on other grounds it must be looked for somewhere in this region.

No. 161.

Kara Agha. Quadrangular cippus in the wall of the Djami. Copy and impression.[1]

ΛΥΡΜΝΗΣΙΘΕΟC
ΕΡΜΟΓΕΝΟΥC
ΤΗΙΔΙΑΓΥΝΕΚΙ
ΔΟΥΔΑΓΛΥΚΥ
ΤΑΤΗΜΝΗΜΗΣ
ΧΑΡΙΝ

[Α]ὐρ. Μνησίθεος
Ἑρμογένους
τῇ ἰδίᾳ γυνεκὶ
Δούδᾳ γλυκυ-
τάτῃ μνήμης
χάριν.

No. 162.

Tchetme. In the wall of the Djami. Panel with mouldings on all sides, and a circular festoon below the inscription. The top moulding has been hewn away. It probably contained the beginning of the inscription. Immediately beneath the top moulding is the following inscription. Copy.

ΓΗCΗΑΝΙΔΟCCΑΥ
ΤΗΚΑΙΤΕΙΜΟΔΑШ
ΥΙШΜΝΗΜΗC
ΧΑΡΙΝ

---

[1] In line 1, NHϹ are in ligature. The sigmas vary as indicated.

. . . .
γήσῃ 'Ανιδος? [ἐ]αυ-
τῇ καὶ Τειμο(λ)άῳ
υἱῷ μνήμης
χάριν.

Leaving Engilli Mr. Haynes passed through the villages Gedil,
Aghayüt, Regiz, Ortakieui, Kotchash, Yazir, Tchetme, meeting me at
Daghan Hissar. On this excursion he copied Nos. 163–173. The
time occupied by Mr. Haynes between Engilli and Daghan Hissar
was 4 h. 41 m. His route lay wholly in the plain, while mine crossed
a succession of low spurs that run off from Sultan Dagh (see the map
as cited above), but very near the point where the mountain ceases
and the plain begins. This accounts for the fact that he made the
trip in less time than I did (5 h. 5 m.), although my route was the
straight line.

No. 163.

*Regiz. In the foundation of an old Turkish building.
Copy and impression by J. H. Haynes.*

ΛΥ▨ΛΟΥΚΙΟΣΛΟΥΚΙ
ΟΥΜΕΝΕΛΛΟΥΙΔΙΑΓΥ
ΝΑΙΚΙΔΟΜΝΗΘΥΓΑΤΡΙΜΕ
ΝΕΜΑΧΟΥΧΑΡΙΔΗΜΟΥΣΕ
ΛΙΝ▨ΕΩΚΑΙΜΑΤΡΙΑΙΛΑ
ΡΑΜΝΗΜΝΗΜΗΣΧΑΡΙΝ

[Α]ὐ[ρ]. Λούκιος Λουκί-
ου Μενελ[ά]ου ἰδίᾳ γυ-
ναικὶ Δόμνῃ θυγατρὶ Με-
νεμάχου Χαριδήμου Σε-
λιν[δ]έω[ς?] καὶ ματρὶ Αἰλά-
ρᾳ μνήμης χάριν.

No. 164.

Regiz. Fragment in the old Turkish building. Copy and impression by J. H. Haynes.[1]

N H P H Γ⸍⸂ Ο ⸍⸍

H P O N B I O N O I K O N Γ⸍⸍⸍

[ὃς ἂν κακουργήσῃ τοῦτο τὸ μνημεῖον?]
[ὀρφανὰ τέκνα λίποι, χ]ῆρον βίον, οἶκον [ἔρημον],
[ἐν πυρὶ πάντα δάμοιτο, κακῶν ὑπὸ χεῖρας ὄλοιτο?].

No. 165.

Kotchash. Apparently a cornice piece; inserted above the lintel of the door of the Djami. The inscription is in one continuous line. Copy and impression by J. H. Haynes.

ΑΝΕΚΕΝСΘΙΟΝΑΟСΙΥΠΕΡΑΓΙΑΘΚΟСΤᴚΚΥΡᴚΘΕΩΔΟΡᴚ
ΚΕΜΑΓΙСΤΡᴚΤᴚΚΑΡΑ͞ΝΔΙΑСΥΝ⚊ΡΟΜΙСΙ͞ΩΟСΤΙΑΡΙᴚΚΕ
ΕΠΙСΚΕͲͲᴚΒΑСΙΛΕΥ͞ΩΝΒΑСΙΛΙᴚ

. . . . ᾽Αγία Θ(εοτό)κος τοῦ κύρου Θεοδόρου
κὲ μαγίστρου τοῦ . . . . ὀστιαρίου κὲ
ἐπισκεπτίτου βασιλευώντων Βασιλίου [κὲ ᾽Ιωάννου?]

Basil and John reigned jointly from 969–976 A.D.
Concerning the name Kotchash, see *Papers of the American School at Athens*, Vol. III. p. 38, footnote.

---

[1] In line 1 HΣ are in ligature.

No. 166.

Kotchash. Altar with human figure in relief in the Djami. Copy and impression by J. H. Haynes.[1]

A Y P H Λ I O C C O Y [uncut]
A Y Ζ A N Ꮽ N Θ O

Αὐρήλιος Σού[σου?]
Αὐξάνων Θ[ευγένους?]

No. 167.

Kotchash. Altar in the vestibule of the Djami. It was so dark that Mr. Haynes could not see to read it, and a violent wind made the impression worthless.[2]

O Π Λ Ọ N O Ϲ Δ
A Π Π A Ϲ O T I P I
5   Θ P E Y A N T I Θ Y Γ A T C
O Ϲ K A   ଠ K I Λ Λ A
Ϲ Y N T Ꮽ
N H M

"Οπλωνος Δ[ιογένους?]

· · · · · · · ·

· · · · · · · ·
'Αππᾶς ὁ Τιρι[δάτου?] . . .
5   θρέψαντι θυγατ[έρα]

· · · · · · · ·
σὺν τῷ . . . .
[μ]νήμ[ης χάριν].

---

[1] Letters distinct.   [2] In line 5 NT are in ligature.

No. 168.

Kotchash. In the wall of the Djami. Copy and impression by J. H. Haynes.[1]

```
Α C Κ Λ Η Π Ι Α Δ Η C
    Τ Ρ Ι C
 ░░Α Ν Τ Α Λ Ε Ο Ν Τ Ι
    Α Π Ι Ο Υ Τ W Ε Α Υ Τ Ο Υ
5 ▨Ι W Κ Α Ι Α Υ Ρ Δ Α Ε Ρ Μ Ο Γ Ε Ν
  ▨C Τ Η Γ Υ Ν Α Ι Κ Ι Α Υ Τ Ο Υ Κ Α Ι Τ
  C Υ Ι Ο Ι C Α Υ Τ W Ν Ι Δ Ι Ο Ι C Δ Ι
  ▨Δ Ε Λ Φ Ο Ι C Α Υ Ρ Π Α Π Ι Α
  ▨Ν Τ Λ Ε Ο Ν Τ Ο C Κ Α Ι Α Υ ▨
10 ▨▨▨▨▨Η Π Ι Α Δ Η Π Α Ν Τ Α /
  ▨▨▨Τ Ο C     Μ Ν Η Μ Η
       [head]  Α Ρ Ι Ν
```

    'Ασκληπιάδης

      τρὶς

    [Π]ανταλέοντι

    'Απίου τῷ ἑαυτοῦ

5 [υ]ἱῷ καὶ Αὐρ. Δᾷ 'Ερμογέν-

   [ου]ς τῇ γυναικὶ αὐτοῦ καὶ τ[οῖ-]

   ς υἱοῖς αὐτῶν ἰδίοις δ[ύο?]

   [ἀδ]ελφοῖς Αὐρ. Παπίᾳ

   [Παν]ταλέοντος καὶ Αὐ[ρ].

10 ['Ασκλη]πιάδῃ Παντα[λ]-

   [έον]τος μνήμη[ς]

      [χ]άριν.

---

[1] In line 3 Mr. Haynes' copy has \ΑΝΤΑ; in line 4 his copy has ΑΠΠΙΟΥ; in line 5 his copy has ᴾΕΙW; in line 6 his copy begins CΤΗ. The changes have been made on the authority of the impression. Ligatures occur: line 6, ΗΓ· line 7, ΑΥ; line 9, ▨ΝΤ, ΝΤ; line 10, ▨ΗΠ, ΗΠ, ΝΤ; line 11, ΝΗ.

No. 169.

Kotchash. In the wall of the Djami. Copy and impression by J. H. Haynes.[1]

ЕРМОГЕΝΗΘΕΑ
ΓΕΝΟΥCΑΝΕCΤ
CΕΝΜΝΗΜΗCΧΑΡΙΝ

['Ο δεῖνα]
'Ερμογένῃ Θεα-
γένους ἀνέστ[η]-
σεν μνήμης χάριν.

No. 170.

Kotchash. On an altar built into the wall of the Djami. Copy by J. H. Haynes.[2]

ΑΛΙΠΕΡШCΑΓΑΕΟΠΟ
ΔΟCΟΥΕΓΝШΥΙШ
ΜΝΗΜΗCΧΑΡΙΝ

'Αλιπέρως 'Αγα[θ]όπο-
δος Οὐέγνῳ? υἱῷ
μνήμης χάριν.

---

[1] Ligatures occur: line 1, NH; line 3, NMNHMHϹ.
[2] Line 3, HM are in ligature.

No. 171.

Kotchash. In the corner of the Djami. Copy by J. H. Haynes.

EPIMANHNAΠ·Π
NEKPATHNAΠ·π

Λ
ΛIΘKE
"O

No. 172.

Kotchash. On the post of a door. Copy by J. H. Haynes.

PIEB
OITV
XN
CKO
ΘNIKE
OV

No. 173.

Kotchash. Altar stone in the Djami. Copy by J. H. Haynes.

AVλ
TIMOΘ
IΔIШAYΘE
AΓENHΜNH
ΜHCXAPIN

Aὐ. [Μάρκος?]
Τιμοθ[έῳ υἱῷ]
ἰδίῳ Aὐ(ρ). Θε-
αγένῃ μνή-
μης χάριν.

Mr. Haynes found ruins at Regiz and especially at Kotchash. These ruins are late.

No. 174.

Daghan Hissar. On a sarcophagus used as a water-trough in a fountain near a Djami. Copy.

░░A Ν ░░░░░░░░░░░░░░N K N Ö I N░
M A Ε I Δ O K C░░░░░░░A Δ Ε I T O Y

The inscription is apparently Phrygian ; see the following inscription, No. 175.

Daghan Hissar is a modern town without any ancient remains.

June 24. Daghan Hissar to Ashagha Dinek, 4 h. 18 m. We again cross Sultan Dagh to the plain of Kara Aghatch, blank on the old maps, but now filled in by my journeys of 1884 and 1885 (see the map in *Papers of the American School*, Vol. III.). As will be seen from the map the road does not begin to cross Sultan Dagh from Daghan Hissar, as given on Tchihatcheff's map. In reality it goes up the gorge, at the mouth of which Kara Agha is situated. We ascend from Tchetme and join the true road in 36 m. The road crosses a real pass ; the ascent on the eastern side is gentle but steady ; the descent on the western side is sharper and more precipitous. The pass is low.

June 25. Ashagha Dinek to Yalowadj, 5 h. 50 m. We pass Oghras to Tcharük Serai.

No. 175.

Tcharük Serai (Ulumahallü). In the wall of a Djami. See my Preliminary Report, *p.* 11. *Copy and impression.*

I O Ɔ N I Ɔ E M O N K N O Y M A
N E K A K O N Δ A K E T A I N I
M A N K A T I E T I T T E T I
K M E N O Ɔ E I T O Y

The inscription is Phrygian ; see *Papers of the American School at Athens*, Vol. III. No. 571.

No. 176.

Tcharük Serai (Ulumahallü). In the wall of a Djami. Copy.

```
ʸOCTOYKIΛΛANI
ⲱIKA·IΓPAMMATEY
ˢⲈAYTⲱIZⲱN▨▨
IMNHMHCⲈNⲈKⲈN
```

οστουκιλλανι
ωι? καὶ γραμμτεὺ
ς ἑαυτῶι ζῶντ-
ι μνήμης ἕνεκεν.

Tcharük Serai lies in a fertile little valley about an hour east of Kara Aghatch. It is a cluster of seven large and prosperous villages. The whole community goes by the name of Tcharük Serai; but each of the seven villages has its own distinctive name, with the addition of the Arabic word Mahallü, "*Quarter*" (*e.g.* Tchukurmahallü, Ulumahallü, etc.), in short, precisely the same state of affairs as prevails in the Yalowadj of to-day, where there are twelve "Quarters." Perhaps here, as at Yalowadj, these "Quarters" represent ancient vici. Tcharük Serai is certainly the site of an ancient town; possibly Pappa must be placed here, but no documentary proof exists at present.

No. 177.

Tcharük Serai (Ulumahallü). Stele with four figures in relief, in two rows of two figures each. Beneath them is the inscription. Copy and impression.

```
ⲐIOYⲐIOYCΛOΛΛIOCΛˑO▨
ΓⲐINOCTITOYYIOCBABⲈ
INTHNⲈAYTOYΓYNAIKA
▨▨NⲈCTHCⲈNⲘNH
ⲘHCXAPIN
```

Θιουθίους Λόλλιος Λο[ν]-
γ(ε)ῖνος, Τίτου υἱὸς, Βαβε-
ὶν, τὴν ἑαυτοῦ γυναῖκα
[ἀ]νέστησεν μνή-
μης χάριν.

The reading is perfectly certain.

Leaving Tcharük Serai we pass Zengiler, and at Man Agha we copy the following inscription.

Nos. 178–181.

Man Agha, a village about twelve miles to the southeastward of Yalowadj. Roman milliarium at the little, or lower, Djami. Copy and impression.[1]

Side *A*, Nos. 1 and 2.

```
        D D N N
  F L C L C O N S T A N T I N O M A X
          I M O
E T F L I V L C O N S T A N T I O E T F L
  C L C O N S T A N T I░░░░R ! ! ! ;
  V I C T O R I B  ʃ E M P A V G G
  Λ B A N T I O C H I A
  I░░░░░░░░░░░░░░░░░░░░░G I░░
  P O N T I F · M A X · T R I B
  P O T  X I I I  C O S  I I I
          P · P
```

MᵀP U

· [1] This was published in my *Preliminary Report on an Archæological Journey made in Asia Minor during the Summer of 1884*, but for some reason, which I cannot explain, several errors have crept into the text as presented there. Above I present a corrected text.

Side *B*, No. 3.

IMPCMAVRVAL
MAXIMIANO
FINVICTAVG
▓▓▓S⁚ON⁝⁝P⁚A▓P
R▓▓▓▓▓▓▓▓
▓▓B▓CAESARIB

Side *B*, No. 4.

      IMPCAS
MARCAPO
ETIMPCAES
MAVRVAL
MAXIMIANO
PFAVG ⟨▓▓

It seems that Side *A* must fall into two fragments of inscriptions: the one having been partially hacked off to give place to a new one, which, in its turn, was left in an incomplete state. It is not easy to unravel Nos. 1 and 2, owing to the fact that No. 2 does not furnish data enough to justify an attempt at restoration. Accordingly I simply transcribe in minuscules the text as it stands on the stone:

Nos. 1 and 2.

(Duobus) d(ominis) n(ostris) Fl(avio) Cl(audio) Con-
stantino Maximo [P(io) F(elici) Inv(icto) Aug(usto)]
et Fl(avio) Jul(io) Constantio et Fl(avio) Cl(audio)
Constanti[no . . . .] victorib(us) semp(er)
[A]ug(ustis)
      [A]b Antiochia
Pontif(ici) Max(imo), trib(uniciae) pot(estatis) XIII,
Co(n)s(uli) III, p(ro) P(raetore). Mi(lia) p(assuum) (quinque).

We are not without proof that the people of this region of country were ignorant both of the number and the names of their rulers; see Mommsen's remarks in the *Ephemeris Epigraphica*, 1884, p. 594, No. 1403. This ignorance is shown in our inscription by the omission of the name of Flavius Julius Constans, and by the fact that the DDNN claims that there were only two emperors. .

No. 3.

[Imp(eratori) Caes(ari) C. Val(erio) Diocletiano
P(io) F(elici) Invict(o) Aug(usto) et] Imp(eratori)
C(aesari) M. Aur(elio) Val(erio) Maximiano [P(io)]
F(elici) Invict(o) Aug(usto) [et Fl(avio) Val(erio)
Constantio et Gal(erio) Val(erio) Maximiano
no]b[ilissimis] Caesarib(us).

No. 4.

The reading of No. 4 is curious, but it is certain, and is vouched
for by the impression.   Lines 3, 4, 5 seem to show that the inscription is a repetition of No. 2, inasmuch as the name must be M. Aur.
Val. Maximianus.   But while we should look for the name of Dio-
cletianus in line 2, we find only strange letters.   It may even be that
No. 4 contains fragments of two inscriptions.

The stone is at least seven miles out of place.

Man Agha is situated in a deep Dere.   Henceforward our road
crosses a succession of low spurs that run off from Sultan Dagh.
Between each of these spurs there is a narrow valley, each with its
village and stream of water.

No. 182.

Örkenez.   Copied by J. H. Haynes.[1]

O Λ Λ H Γ A N M A K A P O O T E
Λ A K H Δ O N O C H Δ I E P O I C
Λ A C T O N K O Λ Λ H Γ O Y
Α P E T H C O A Λ O C A Θ A N A T
5   Ϊ O N N E O N A M B P O C I O N M H
Η C Y N E Φ Y N E Π A Λ A I O I C
H T H P Δ E Ш Δ E I N Δ Y C A P I
O T O K E I A T E M A Γ N A
Λ E K I T H K A C E I T E M E Λ O C
10   I A P K A T E E T O H Δ Y
O N C O Φ O N i H T P O N E I K O C
Γ O N Δ I O N T A E I O C

---

[1] In line 5 ΝΝΗ are in ligature.   In line 11 end the C is written above the O.

Continuing our march towards Yalowadj we passed Kuyudjak, where we copied No. 105, and Gemen, where we copied Nos. 107 and 116. These inscriptions belong properly to Yalowadj, and for that reason they have been placed among the inscriptions of that place.

June 26. Yalowadj to Man Agha, 4 h. 15 m.

June 27. Man Agha, *via* Tchartik Serai and Kara Aghatch, to Beikieui, 3 h. 11 m. At Kara Aghatch the following inscriptions were copied.

No. 183.

Kara Aghatch. Epistyle block ornamented with an eggstaff in the door of the Khan. Length, 1.72 m. The inscription occupies the whole length of the block, and was continued on other blocks to the right and left of this one. Copy.

THΛΑΕΜΑΧΟΣΕΡΜΟΓΕΝΟΥΣΤΟΥΤΟΥΛΟΥΡΑΣΕΩΣ

ΤΥΛΟΙΣΕΚΤΩΝΙΔΙΩΝΔΙΟΣΚΟΡΟΙΣΘΕΟΙΣΕΠΗΚΟΟΙΣ

. . . Τηλέμαχος Ἑρμογένους ⟨τοῦ⟩ Τουλουράσεως . . . . . .
. . . σ]τύλοις ἐκ τῶν ἰδίων Διοσκόροις θεοῖς ἐπηκόοις . . . . . .

The names Τηλέμαχος, Ἑρμογένης, and Τουλουράσεως seem to have been common in this region, *e.g.* see *Papers of the American School,* Vol. III. Nos. 323, 328, 329. In the last (No. 329) a sister of Τηλέμαχος seems to be mentioned : Βαβὶς Ἑρμογένους Τουλουράσεως. This last is a queer name, and yet we have analogical formations from this very region. Thus in the *Papers of the American School,* Vol. III. No. 329, we find Τουλλανδός, and in No. 330 Τουραμμᾶς; while in No. 323 we read Τολουράσις, which is probably a mistake on the part of the stonecutter for Τουλουράσις.

No. 184.

Kara Aghatch.   Large stone partly buried in the foundation of a Minare.   Owing to the opposition of the natives I could not unearth it.   Copy.

```
ΜΕΝΕΛΑΟϹΜΕΝΕ
ΟϹΑΥΡΜΕΝΕϹΘΕѠ
ΙΝΕΛΑΟΥΥΙΟϹΕΠΕ
ΓΡΥΑΟΥΚΗΜΗΝΕΙΕ
ΝΟΜΗΝΟΥΚΕΙΜΙΟΥ
ΜΕΛΕΓΜΟΙΥΓΙΑΙΝΕ
ΠΑΡΟΔΕΙΤΑ
ΜΗΤΕΟΙΚΗϹΙϹΠΕΡΙΛϹΦΘΗϹ
ΤΑϹΤΟΥϹΘΕΟΥϹΚΕΧϹΛΟΜΕΝΟ
Τ        ΝΟ    ΙΤΟΙΚΟΥ
```

Μενέλαος Μενε[λάου . . .]
ος Αὐρ. Μενεσθέω[ς . . .]
[Με]νελάου υἱὸς, ἐπέ- . . .
γρ[α]ψα? . . . .
.   .   .   .
.   .   .   .   . ὑγίαινε . . . .
παροδεῖτα . . . .
μήτε οἴκησις . . . .
τας τοὺς θεοὺς [κ]εχ[ο]λομένο[υς] . . .
.   .   .   .

No. 185.

Kara Aghatch. Large stone nearly buried in the foundation of the Djami. The insolence of the mob so disturbed and vexed me that I did not have the bottom of the stone unearthed. I think the inscription will be found to be complete. See Mittheilungen des Deutschen Archaeologischen Instituts in Athen, 1883, p. 75. Copy.

```
TOHPWEIONAYP
MENECΘEWCMENEΛA
OTOYAΓYKIOYAΛTAΔE
WCBOYΛEYTOYTHCTEΠAΠΠO
```

```
THCΓYNAIKOCAY
TOYKATECKEYACAI
```

Τὸ ἡρῷον Αὐρ.
Μενεσθέως Μενελά-
ο(υ) τοῦ Λ[ο]υκίου Ἀλταδέ-
ως βουλευτοῦ τῆς τε Πάπ[πη-?]

τῆς γυναικὸς αὐ-
τοῦ κατεσκεύασα[ν]
[τε τὸ ἡρῷον ἑαυτοῖς
καὶ τοῖς ἰδίοις].

Var. Lect.

Line 2, MЄNЄCO, *Mittheilungen;* line 3, OΓOYΛOY in initio, ΔC in fine, *Mittheilungen;* line 4, TCτιλπο in fine, *Mittheilungen;* line 5, C omitted by *Mittheilungen;* line 7, I added in fine by *Mittheilungen;* after line 7 the *Mittheilungen* give two lines not copied by me on account of the rudeness of the mob, viz. : —

ΤЄΤΟΗΡШЄΙΟΝЄΑΥΤΟ

ΚΑΙΤΟΙCΙΔΙΟΙC

. . . .

No. 186.

Aïpler, properly Eyuplar, *see* Papers of the American School, *Vol. III., footnote to No.* 394. *This is a second Eyuplar. Fragment in a wall. See my* Preliminary Report, *p.* 11. *Copy.*

ΙΟCΚЄCЄΜΟΝΤΟΚΑΚΟΝΟ *L%*

June 28. Derekieui to Munafer, 5 h. 11 m. Shortly after passing Tchaltü we enter the pass between Kizil Dagh and the outliers of Sultan Dagh, reaching the watershed in 1 h. 14 m. Leaving Felle (see *Papers of the American School,* Vol. III. pp. 192-193) to our right, we reach the ruins called Monastir in 2 h. 30 m. from Beikieui. Monastir is situated at the northern limit of the plain of Kirili Kassaba ; the ruins are unimportant. Here Mr. Ramsay locates Misthia (see *American Journal of Archaeology,* I. p. 146). Near the site there is a mound, which may have served as an Acropolis. It has every appearance of an artificial mound. Leaving Monastir we traverse the plain, reaching Kirili Cassaba in 56 m.

No. 187.

Kirili Cassaba. On a sarcophagus in a street. Mittheilungen des Deutschen Archaeologischen Instituts in Athen, 1883, *p. 77. Copy.*

ₚ꜀ᵞΛΙΟϹΜΑΡΚΟϹϹΤΑΤΙШΝΑΡΙΟϹ
ΑΘΛΑΙΑΜΑΤΡШΝΗΤΗΚΑΙΘΛΠΙΔΙ
ϹΥΝΒΙШΓΛΥΚΥΤΑΤΗΜΝΗΜΗϹ
ΧΑΡΙΝ

['Ιού]λιος Μάρκος στατιωνάριος
[Λο]λ[λ]ία Ματρώνη τῇ καὶ 'Ελπίδι
συνβίῳ γλυκυτάτῃ μνήμης
χάριν.

Concerning the *Stationarii*, see the note in the *Mittheilungen*, as cited above.

Var. Lect.

Line 1, *Mittheilungen* has ΛΙΟϹ, and indicates no break in init.; *Mittheilungen* reads ΡΚϹ, and indicates a lacuna after these letters. Line 2, *Mittheilungen* reads ΑΟΛΛΙΑ in init. and ΓΙΔΙ in fine.

No. 188.

Kirili Cassaba. In a panel on a sarcophagus. Bulletin de Correspondance Hellénique, 1886, *p.* 502. *Copy.*

ΘΕΟΦΙΛΟϹϹΕ
ΒΑϹΤΟΥΑΠΕΛΕΥΘΕΡΟΥ
· ΕΠΙΤΡΟΠΟϹ
ΚΑΛΛΙΓΕΝΕΙΘΡΕ
ΠΤШΤΕΙΜΙШΤΑΤШ
ΜΝΗΜΗϹΧΑΡΙΝ

Θεόφι[λ]ος Σε-
βαστοῦ ἀπελεύθερο[ς],
ἐπίτροπος
Καλλιγένει θρε-
πτῷ τειμιωτάτῳ
μνήμης χάριν.

No. 189.

Kirili Cassaba. Stele now in the Christian cemetery. Copy.

OYNOYCIΔIAN	. . . . . . ουνους ἰδίαν
OYΛΛΕΝΤΙΛΛΑΝ	Οὐ[a]λέντιλλαν
THNAΞIOΛOΓWTA	τὴν ἀξιολογωτά-
THNMMATPWNANCYN	την ματρώναν συν-
ΓΕΝΙΔΑCYNKΛHTI	γενίδα συνκλητι-
KWNTHNCEMNOTATHN	κῶν τὴν σεμνοτάτην
KAIΦIΛOTEKNONΓYNA	καὶ φιλότεκνον γυνα[ῖ]-
KAKAΛΠOYPNIOY	κα Καλπουρνίου
MAPKEΛΛOYTOYKPA	Μαρκέλλου τοῦ κρα-
TICTOY	τίστου.

The name Οὐαλέντιλλα occurs in an inscription of Antiochia Pisidiae above, No. 138.

Kirili Kassaba is a small market town, as the name indicates. It is very malarious during the summer and early fall. Leaving Kirili Kassaba we traverse a rolling country, passing Tchukurkend and Yenidje, and reaching Munafer in 1 h. 45 m. Munafer is a wretched village situated on the edge of a swamp, and like all the villages near the lake it is very unhealthy. On the contrary, the numerous and prosperous villages on the high ground along the northeastern edge of the plain of Kirili Kassaba are not malarious.

June 30. Munafer, *via* Eflatun Puñar, to Yonüslar, 7 h. 15 m. We visit Eflatun Puñar in order to secure photographs of the important monument. One of these photographs has been published by Dr. William Hayes Ward in the *American Journal of Archaeology*, Vol. II. (1886) pp. 47–51 and Plate I. Professor Kiepert thinks that the name of the place is Eflatun, not Elflatun. Eflatun is the Arabo-Turkish pronunciation of Platon, a name which cannot have the Arabic article *el*.

Leaving Eflatun Puñar we pass Selki, immediately beyond which place we ascend to a table-land. As we advance, this plateau gradually becomes rolling and barren. The few inscriptions of Yonüslar are given in the *Papers of the American School*, Vol. III. Nos. 313–315.

July 1. Yontislar to Kizil Ören, 3 h. 18 m. The road leads up the Bagharzik Dere, of old infested by brigands, the terror of the caravans between Isparta-Yalowadj and Konia. The mountains on either side are low, but rough and jagged, and abounding in secure hiding-places for robbers. In 2 h. 5 m. the plain of Kizil Ören is reached. Nearly an hour west of Kizil Ören are the remains of a Seldjuk Khan and Djami.

No. 190.

Kizil Ören. On an epistyle block in the cemetery. Letters very faint and blurred. Copy.

HKETONΔOYΛOONTOYΘOΤΕ҉ΛΗΤΗΝOΤONCINCIC
CITONCIONTOCIΓHOΘΕΟΔOPONIINHCITPHAHOCIM
KΕBOHΘHTONΔOYΛ\ONCOYΛΕONTEINA҉÷CTOPCIN
MΕTCITHCOΔHCITOYCIM҉ΦKΓ҉OΔOMOCCYΘΕ

βοήθ]η K(ύρι)ε τὸν δοῦλον τοῦ θ(εου) . . . .
. . . . Θεόδορον . . . . . .
K(ύρι)ε βοήθη τὸν δοῦλόν σου Λεοντίνα . . . . . .
. . . .

July 2. Kizil Ören to Konia, 6 h. 20 m. The road is uninhabited and monotonous, lying mostly in ravines and defiles. In 1 h. 5 m. from Kizil Ören we pass the ruins of a Seldjuk Khan. In 2 h. 30 m. from this Khan we reach another and better preserved Khan, situated in a little valley just at the point where our road unites with the horse road from Ak Shehir to Konia. This succession of Seldjuk Khans—first between Egherdir and

Gelendos, then near Kizil Ören, and finally the two between Kizil Ören and Konia—shows that we have traversed the great commercial road between the Seldjuk Konia and the seaboard. Leaving this Khan we cross a mountain, and in 1 h. 46 m. we reach the western limit of the great plain of Konia.

No. 191.

Konia (Iconium). *Slab now forming the Musalla Tashii in the southern cemetery. My copy. (I was also furnished with a faulty copy by Dr. S. N. Diamantides.) Length, 2.50 m.; height, 0.60 m.*

AIΛIATATATONTEPIBOΛOONENAPXΘENTAⁱⁱⁱⁱⁱⁿYOAPPOYNⁿⁿOYΔIOMHΔOYCTOY
ANΔPOCAYTHCTEΛECACAKATAKEIMENOYTOYΔIOMHΔOYCKAITOYYIOYAYTWNΔOMNOY
ENAYTWAΦWCIWCENKAIEICTHNEAYTHCKHΔEIANOYΔENIΔEEEECTAI
ETTEICBIACACΘAIHETTEICENENKEINTTTWMAEANΔETICEICBIACHTAI
YTTOKEICETAITWIEPWTATWTAMEIWⁱMYPIOICTTENTAKICXEIΛIOIC

Αἰλία Τάτα τὸν περίβολον ἐναρχθέντα [ὑπ]ὸ Ἀρρουν[τί]ου Διομήδους τοῦ ἀνδρὸς αὐτῆς τελέσασα, κατακειμένου τοῦ Διομήδους καὶ τοῦ υἱοῦ αὐτῶν Δόμνου ἐν αὐτῷ, ἀφωσίωσεν καὶ εἰς τὴν ἑαυτῆς κηδείαν. οὐδενὶ δὲ ἐξέσται ἐπεισβιάσασθαι ἢ ἐπεισενέκειν πτῶμα. ἐὰν δέ τις εἰσβιάσηται ὑποκείσεται τῷ ἱερωτάτῳ ταμείῳ (δηναρίοις) μυρίοις πεντακισχειλίοις.

No. 192.

Konia. Huge stone near the outskirts of the city as one journeys towards Ak Serai. Copy.

```
ЄΒΟΥΡΗΝΛΝ·ΜΑΖΙ
ΜΑΝ☒ΘΥΓΑΤΕΡΛΓΑ
ΙΟΥ·ЄΒΟΥΡΗΝΟΥ☒
ΟΥΑΛΕΝΤΟC☒ΓΥΝ☒
☒ΚΑ·ΚΟΙΝΤΟΥ·ЄΡ☒
☒☒ΛΝΟΥ·ΜΑΖΙ☒☒
```

['Ε]βουρήν[α]ν Μαξί-
[μ]αν θυγατέρ[α] Γα-
ίου 'Εβουρήνου
Οὐάλεντος, γυν[α-]
[ἰ]κα Κοίντου 'Ε[β-]
[ουρήνου Μαξίμου

No. 193.

Konia. Stele recently found in a part of the city walls which have been demolished for building purposes. Copy.

```
ΑΥΡ·ЄΥΤΥΧΙΟΥ
ΖШΤΙΚΟΥCΥΝΤΗ.C
ΥΝΒΙШΑΝΤШΝΙΑ
ΑΝЄCΤΗCΑΜЄΝ
ΖШΝΤЄCЄΑΥΤΟΙC
ΤΟΝΤΙΤΛΟΝΜΝΗ
ΜΗCΧΑΡΙΝ†
```

Αὐρ. Εὐτυχίου
Ζωτικοῦ σὺν τῇ σ-
υνβίῳ ᾽Αντωνίᾳ.
ἀνεστήσαμεν
ζῶντες ἑαυτοῖς
τὸν τίτλον μνή-
μης χάριν.

No. 194.

Konia.   Slab from the recently demolished walls.   Copy.

```
Ο Υ Α Δ Ο Υ C Κ
Α Ι Δ Ο Υ Δ Α Π
Ρ Ο Κ Λ Ш Τ Ш
Α Δ Є Λ Φ Ш Μ
Ν Η Μ Η C Χ Α Ρ
```

Οὐάδους κ-
αὶ Δούδα Π-
ρόκλῳ τῷ
ἀδελφῷ μ-
νήμης χάρ(ιν).

No. 195.

*Konia.   Stele with reliefs built into the wall of the Djami
Sultan Aladdin.   My copy.   (A faulty copy was also furnished me by Dr. S. N. Diamantides.)   1.0 m. × 0.55 m.*[1]

---

[1] Ligatures occur: lines 4, TH; 6, MNHM, HC.

ΚΟΙΝΤΟΣΦΟΥ
ΛΒΙΟΣΑΛΕΞΑ░
ΔΡΟ░░░ΩΝΕ
ΑΥΤΩΚΑΙΤΗ
ΓΥΝΑΙΚΙΠΑΥ
ΛΕΙΝΗΜΝΗΜΗΣ
ΧΑΡΙΝ

Κόϊντος Φού-
λβιος Ἀλέ[ξ]α[ν-]
δρο[ς ζ]ῶν ἑ-
αυτῷ καὶ τῇ
γυναικὶ Παυ-
λείνῃ μνήμης
χάριν.

No. 196.

Konia. Copy of Dr. S. N. Diamantides. In the water reservoir of Petros Pappazian. 1.50 m. × 0.55 m.

ΟΝΗCΙΜΟCΚΑΙ
CΑΡΟCΒΑΒΙΓΥΝΑ
ΙΚΙΤΗΝCΤΗΛΗΝ
ΕCΤΗCΕΚΑΙΕΠΕ
ΓΡΑΨΕΜΝΗΜΗC
ΧΑΡΙΝ

Ὀνήσιμος Καί-
σαρος Βαβῖ γυνα-
ικὶ τὴν στήλην
ἔστησε καὶ ἐπέ-
γραψε μνήμης
χάριν.

No. 197.

Konia. Copied by Dr. S. N. Diamantides. 0.85 *m.* × 0.68 *m.*

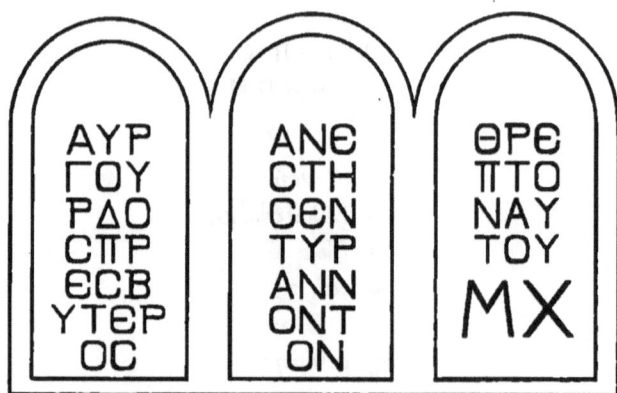

```
ΑΥΡ      ΑΝΘ      ΘΡΘ
ΓΟΥ      ϹΤΗ      ΠΤΟ
ΡΔΟ      ϹΘΝ      ΝΑΥ
ϹΠΡ      ΤΥΡ      ΤΟΥ
ΘϹΒ      ΑΝΝ      ΜΧ
ΥΤΘΡ     ΟΝΤ
ΟϹ       ΟΝ
```

Αὐρ. Γοῦρδος πρεσβύτερος ἀνέστησεν Τύραννον τὸν
θρεπτὸν αὐτοῦ μ(νήμης) χ(άριν).

No. 198.

*Konia. Stele representing the façade of a temple; in the
temple to the left is a standing figure, to the right a seated
figure. Between and above them are various implements
and vessels: bucket with fruit, basket, comb, flowers.
Photograph. Length, 0.395 m.; width, 0.40 m. Copy.*[1]

Α'''ϊ'ˣΩΝΙΑΝΟϹΛΟΝΓΕΙΝ(∅∅∅∅
ΑΘΟΥΓΑΤΡΙΑΥΤΟΥΜΝΗΜΗϹ
ΧΑΡΙΝ

ʼΑ[ντ]ωνιανὸς Λονγείν[ου Τατί- ?]
ᾳ τῇ θυγατρὶ αὐτοῦ μνήμης
χάριν.

---

[1] The second symbol in line 1 is probably a ligature for NT; in line 2, OY are
in ligature.

No. 199.

Konia.   In a street.   My copy.   (A faulty copy was also furnished by Dr. S. N. Diamantides.)   10 *m.* × 0.50 *m.*

```
A I Λ I A K A I C I A A T T A
⧫\ Ш A N Δ P I T O N B Ш
M O N A N E C T H C E
Φ I Λ A N Δ P I A C
    X A P I N
```

Αἰλία Καισία ᾿Αττά-
[λ]ῳ ἀνδρὶ τὸν βω-
μὸν ἀνέστησε
φιλανδρίας
    χάριν.

No. 200.

Konia.   Copy by Dr. S. N. Diamantides.   In the wall of a street leading to Sille.   0.80 *m.* × 0.60 *m.*

```
      O Y E C T I N T O C H M A T O Y T O H
      T E Ш C O P Θ O Δ O Ξ O Y K E H Θ O Y C X H
      T I T O C B I O Y Δ I A K O N O C A Π O Y Γ
      Θ O Λ I K H C E K A H O N C K O T A C T H
      Ξ E Y Γ E N E Θ E I C Y Π O Θ E O Y Δ I
        Y T O C K E H Ш C M N H M I O N E N
          K E Y M C E N K E T H E
            T I T Λ O N E Y Π O I E I
            A Ξ E N E K E N
```

N ᐯ Ш

. . . . .]ού ἐστιν τὸ σῆμα τοῦτο· η-
. . . . .]τεως ὀρθοδόξου κὲ ἤθους χη-
. . . . .] Τίτος βίου διάκονος ἀπὸ ὑγ-
. . . . κα]θολικῆς ἐκ[λ]η[σία]ς κ[ὲ] τὰς τη-
. . . . .]ξευγενεθεις? ὑπὸ θεοῦ δι-
. . . . a]ὐτὸς κὲ ἡ ὡς μνημῖον ἐν-
[τάφη . . . . κατεσ]κεύ[α]σεν κὲ τῇ [σ-]
[υνβίῳ . . . .]τὶτλον ἐ⟨υ⟩ποίει
. . . . [εὐνοί]α[ς] ἕνεκεν.

No. 201.

Konia. Copy of Dr. S. N. Diamantides. In the house of A. Koskinides. 0.60 m. × 0.45 m.

```
///////////////////////////////M O
///////Λ Ε Ν Ε Δ Η Μ Ο C Α Ν Π//////
//////N Ε Δ Η Μ////H///Ψ I Υ I C//////
//////H Λ H N Ε Υ///Φ I A C//////////
/////Ε Κ Ε N
```

. . . . μο
[Μ]ενέδημος Ἀν[τιόχου?]
[Με]νεδήμ[ῳ] [ἀνε]ψι[ῷ ἀν]-
[έστησεν τὴν στ]ήλην εὐ[νο]ίας
[ἕν]εκεν.

No. 202.

Sille, near Konia. Copy of Dr. S. N. Diamantides. The stone is now in the church of the Taxiarchs in Sille, but is said to have been brought from Ladik. Length, 1.0 m.; height, 0.35 m.[1]

---

[1] Line 5, HⳐ are in ligature.

ΔΑΗCΑΛΕΖΑΝ
ΔΡΟΥΤΑΤΑΔΗΓΥΝΑ
ΙΚΙΚΕΔΟΔΑΔ▨▨ΠΕΝ
ΘΕΡΑΖШCΗΜΝΗ
ΜΗΓΧΑΡΙΝ

Δάης 'Αλεξάν-
δρου Τάτα δῆ γυνα-
ικὶ κὲ Δόδα δ[ῇ] πεν-
θερᾷ ζώσῃ μνή-
μης χάριν.

The interchange of δ for τ is very interesting.

No. 203.

Konia. Quadrangular cippus (0.42 m. × 0.28 m.) in the house of A. Koskinides. My copy. (A copy was also furnished me by Dr. S. N. Diamantides.)

ΜΗΝΑCΚΑΙΠΕΡCΕΥ
CΠΟCΕΙ
ΔШΝΙ
ΕΥΧΗΝ

Μηνᾶς καὶ Περσεὺς
Ποσειδῶνι εὐχήν.

No. 204.

Konia. Copy of Dr. S. N. Diamantides. In the house of A. Koskinides. 0.60 m. × 0.27 m.

▨ΜΟΙΕΝΗCΕΛ
ΟΡШΤΟΝΤΟΠΝΟ
ΑΙΛΙШΚΥΙΝΠΑΝШ
ΜΕΝΕΔΗΜШ

5 ΚΕΓΥΝΑΙΚΙΑΥΤΟΥΑΙ
ΛΙΑϹΤΡΑΤΟΝΕΙΚΗ
ΑΝΕϹΤΗϹΑΤΟΝΒШ
ΜΟΝΜΝΗΜΗϹΧΑ
ˣΑΡΙΝ
10 ΛΟΥΚΙΑΝΟϹΕΧΕ
ΝΤΟΠΟΝΟΠΙϹШΤΙ
▨ШΜΟΥΜΕΑΝ▨
▨ΤΕΡϹΕΠΙϹΒΙΑ
Φ▨

. . . .
. . . . . . τὸν τόπ[ον]?
Αἰλίῳ Κυιν[τι]ανῷ
Μενεδήμῳ
5 κὲ γυναικὶ αὐτοῦ Αἰ-
λίᾳ Στρατονείκῃ
ἀνέστησα τὸν βω-
μὸν μνήμης ⟨χά-⟩
χάριν·
10 Λουκιανὸς ε(ἶ)χε-
ν? τόπον ὀπίσω . . .
. . . . . ἐὰν [δέ τις]
[ἔ]τερ[ος] ἐπεσβιά[σηται δώσει τῷ κυριακῷ φίσκῳ* κτλ.].

No. 206.

Konia. Small sarcophagus in the court of a house. The two ends represent temples with gables, between whose columns stand figures in bas-relief. On one side are figures of a man and woman, around whom twines a large vine with hanging grapes. Length, 1.12 *m.; width,* 0.41 *m.;*

height, 0.67 m. Photograph and copy. On one side is inscription A, of which I have copy and impression.[1]

A.

```
Τ Ρ Ο Κ Ο Ν Δ Α C
Κ Ε Ο.Υ Α Ν Γ Δ Ι
Β Α C C Ι Ν Τ Α
Θ Ρ Ε Π Τ Α
```

On the other side is inscription B.

B.

```
Κ Ε Α Ν Τ ω Ν Ι Ο Ν Κ Ε Α Π Ι Α▨▨▨Ν Α Δ Ε 𝒦ΟΥ C Κ Ε Α ι Τ Λ▨
Κ Ι Λ Ι C Τ Ρ Ε Α Φ Ι Λ Ο Ν▨▨▨ΤΟΥ Α Ν Δ Ρ Ο C Λ Ο Ν▨▨▨
Μ Ν Η Μ Η C Χ Α Ρ Ι Ν
```

A.

Τροκόνδας
κὲ Οὐάνγδι
Βασσὶν τᾷ
θρεπτᾷ.

B.

['Η δεῖνα ἀνέστησε τὸν δεῖνα]
κὲ 'Αντώνιον κὲ 'Απια[νὸν] ἀδε[λφο]ὺς κὲ 'Α[τ]π[α]λον
Κιλιστρέα φίλον τοῦ ἀνδρὸς
μνήμης χαριν.

An inscription similar to *B* has been published in the *Bulletin de Correspondance Hellénique*, 1883, p. 314, but it must be a duplicate of this one, else inscription *A* would have been published there also.

Concerning the town *Kilistra*, see *Bulletin* as cited, and *Papers of the American School at Athens*, Vol. III. p. 159.

---

[1] Ligatures occur: in *A*, line 2, ΝΓ, ΚΕ. In *B*, line 1, ΚΕ ter, ΝΚΕ; line 3, ΝΗ. In *B*, line 2 end, ΛΟΝ belongs to end of line 1.

The name Τροκόνδας occurs *C.I.G.* 904, 3366 *k*; *Bulletin,* 1879, p. 344, No. 23 (from Isparta), 1883, p. 268, No. 10 (from Cretopolis); Μουσεῖον καὶ Βιβλιοθήκη τῆς Εὐαγγελικῆς Σχολῆς, 1875, p. 129, No. 44 (from Isparta). The name seems to be confined to Pisidia and Lycaonia.

No. 207.

Konia.   Copy of Dr. S. N. Diamantides.   In the prison.
0.80 *m.* x 0.80 *m.*

```
ΑΥΡΗΡΑΚΛΙΑ ˢ Α Ν Ε C
Τ Η C Α Τ Ω Γ Λ Υ Κ Υ Τ Α Τ Ω
Μ Ο Υ Α Ν Δ Ρ Ι Α Υ Ρ Α Ν Ε Ν Κ Λ Η Τ
Ω Δ Ο Μ Ν Ο Υ C Υ Ν Τ Ω Ν Υ
Ι Ε Ι Ω Ν Μ Ο Υ Μ Α Γ Ι Ω Κ
Ε Γ Α Ε Ι Ω - - - ' Ν Τ Ε C
Μ Ν Η Μ Η C Χ Α Ρ Ι Ν
```

Αὐρ. Ἡρακλία ἀνέσ-
τησα τῷ γλυκυτάτῳ
μου ἀνδρὶ Αὐρ. Ἀνενκλήτ-
ῳ Δόμνου σὺν τῶν υ-
ἱειῶν (= υἱῶν) μου Μαγίῳ κ-
ὲ Γαείῳ [ζῶ]ντες
μνήμης χάριν.

Notwithstanding her name it is clear that Heraclia did not speak Greek as her mother tongue. Σὺν with the genitive is interesting enough in itself, but barbarism can go no farther than to give σὺν the genitive, dative, and nominative all at the same time.

No. 208.

Konia.   In the floor of the water reservoir of the Konak. My copy.   (A copy was also furnished by Dr. S. N. Dia-mantides.)   Length, 0.80 *m.; width,* 0.86 *m.*[1]

---

[1] Ligatures occur: line 1, ME, NN; line 2, KE; line 3, HE, KE, MHN; line 4, TH; line 5, MNH.

MENNEACΠPOCBYTEPOC
ΦPOYΓIOYKEΔOMNOCKAIAP
ECTIΔHE.KEMHNΠIΔOCANEC
THCANTWYΔIWAΔEΛΠW
ATTAΠPECBYTEPWMNHM
HCKAPIN

Μεννέας προσβύτερος
Φρουγίου κὲ Δόμνος καὶ ᾿Αρ-
εστίδης κὲ Μήνπιλος ἀνέσ-
τησαν τῷ ὑδίῳ ἀδελπῷ
῎Αττᾳ πρεσβυτέρῳ μι·ήμ-
ης κάριν.

Nos. 209–210.

Konia. Copy of Dr. S. N. Diamantides. On the road to Karaman, and in the bridge called Tcharshü Abba. Length, 1.45 m.; height, 0.80 m.

A.	*B.*
TABEICKOM	TABEICEK
HCENTONA	OCMHCEN
ΔEΛΦONAY	THNAΔEΛ
TOYTAPACI	ΦHNATIA
NTONE.ΠEIK	NINTHNΔI
H H▨▨AIAΓNO	AKONICCA
M X	N M·X·

A.

Ταβεὶς [ἐ]κό(σ)μ-
ησεν τὸν ἀ-
δελφὸν αὐ-
τοῦ Ταράσι-
ν τὸν ἐπ(ι)εικ-
ῆ [κ]αὶ ἀγνὸ[ν]
μ(νήμης) χ(άριν).

B.

Ταβεὶς ἐκ-
όσμησεν
τὴν ἀδελ-
φὴν ᾿Ατια-
νὶν τὴν δι-
ακόνισσα-
ν μ(νήμης) χ(άριν).

No. 211.

Konia.　Tetragonal cippus now in the house of Dr.
Diamantides, copied by J. R. S. S.

```
   Μ Ο Υ Λ Π Ι Ο C Η Ρ▨
   Κ Λ Ε Ι Τ Ο C Ε Α Υ▨▨
   Κ Α Ι Κ Λ Α Υ Δ Ι Α Γ Υ Ν
   ▨ Ι Κ Ι Α Υ Τ Ο Υ Κ Α Ι Τ Ε
 5 Κ Ν Ο Ι C Α Υ Τ Ω Ν Τ Η Ν
   Λ Α Ρ Ν Α Κ Α Κ Α Ι Τ Ο Ν Β▨
   Μ Ο Ν Α Λ Λ Ω Δ Ε Μ Η Θ
   Ε Ι Ν Α Ι Ο C Δ Ε Α Ν Ε
   Π Ι C Β Ι Α C Η Τ Α Ι Η Α
10 ▨▨▨ Ι Κ Η C Ε Ι Ε Χ Ο Ι Τ Ο
   ▨▨▨ Η Ν Α Κ Α Τ Α Χ Θ Ο Ν Ι
   ▨ Ο Ν Κ Ε Χ Ο Λ Ω Μ Ε
          Ν Ο Ν
```

Μ. Οὔλπιος Ἡρ[ά]-
κλειτος ἑαυ[τῷ]
καὶ Κλαυδίᾳ γυν-
[α]ικὶ αὐτοῦ καὶ τέ-
5　κνοις αὐτῶν τὴν
[λ]άρνακα καὶ τὸν β[ω]-
μὸν· ἄλλῳ δὲ μὴ θ-
εῖναι· ὃς δ' ἐὰν ἐ-
[π]ισβιάσηται ἢ ἀ-
10　[δ]ικήσει ἔχοι τὸ-
ν Μ]ῆνα Καταχθόνι-
[ο]ν κεχολωμέ-
νον.

No. 212.

Konia. Copied by Dr. S. N. Diamantides: Ἐπὶ πέτρας ἐστρωμένης ἐν ταῖς βαθμῖσιν Ἀρμενικῆς τινος οἰκίας· ὕψ. 0.45, πλ. 0.60, ὕψος γραμμάτων 0.04.[1]

```
Π Υ Λ Α Δ Η C
Κ Α Ι Α Ι Λ Ι Α Ζ ο Η
Η Γ Υ̇ Ν Η Α Υ Τ Ο Υ
Ζ Ω Ν Τ Ε C Ε Α Υ Τ ο Ι C
Ε Π Ο Ι Η C Α Ν Τ Η Ν
Λ Α Ρ Ν Α Κ Α Κ Α
//////Τ Ν Ι Γ Τ//////
```

Πυλάδης ·
καὶ Αἰλία Ζοὴ
ἡ γυνὴ αὐτοῦ
ζῶντες [ἑ]αυτοῖς
ἐποίησαν τὴν
λάρνακα κα[ὶ]
[τὸν τί]τ[λον].

No. 213.

Konia. Copied by Dr. S. N. Diamantides: Ἐπὶ τεμαχίου πέτρας εὑρισκομένης ἐν ταῖς βαθμῖσι ὀθωμανικῆς τινος οἰκίας.[2]

```
Ο C Κ Α Ι Α Θ//////////
Ω Ν Κ Ε Β Α C Ι//////////
Ε Ν Η Α Υ Τ Ο Υ Α Ν Ε C
Τ Η C Α Ν C Τ Η Λ Η Ν
Β Ε Ν Η Θ Ω Τ Ε Κ · Λ Α
```

---

[1] Ligatures occur: line 3, ΗΓ, ΝΗ; line 4, ΤΕ; line 5, ΗC, ΝΤΗΝ.
[2] Ligatures occur: line 2, ΚΕ; line 5, ΝΗ.

. . . .
ος καὶ 'Αθ[ηνί-]
ων κὲ Βασὶ[ς ἡ γ-]
ενὴ αὐτοῦ ἀνέσ-
τησαν στήλην
Βενηθῳ? τέκ[νῳ?] . . .

No. 214.

Konia.   Copied by Dr. S. N. Diamantides: 'Επὶ μαρμάρου
εὑρισκομένου ἐντὸς τοῦ νεκροταφείου τῆς 'Εκκλησίας Μετα-
μορφώσεως· ὕψ. 0.95, πλ. 0.40, παχ. 0.18.[1]

```
MIPOCMONA
ZШNKENEC·
ШPANECTHC
AMENTШГΛΥ
KYTATШHMШN
ΠATPIEYCEBIШ
KETHMHTPIHM
ШNΠШMHMN
HMHCXAPIN
```

Μίρος Μονᾶ
ζῶν κὲ Νέσ[τ-]
ωρ ἀνεστήσ-
αμεν τῷ γλυ-
κυτάτῳ ἡμῶν
πατρὶ Εὐσεβίῳ
κὲ τῇ μητρὶ ἡμ-
ῶν Πώμῃ μν-
ήμης χάριν.

---

[1] In line 8 MN are in ligature.

No. 215.

Konia. Copied by Dr. S. N. Diamantides: Ἐπὶ πέτρας
εὑρισκομένης ἐν Ἰκονίῳ.

ΑΥΡΙΛΙΟCΜΑΡΚΟC
ΚΕCΙCΙΝΟCΔΙΑΚШΝ
ΚΕΑΛΕΖΑΝΔΡΟC
ΑΝΕCΤΗCΑΜΕΝΤΟ
ΝΤΙΤΛΟΝΤΟΥΤΟΝ
ΔΟΥΜΕΤΑШΠΡΕCΒΥΤΕΡШ
ΜΝΗΜΗCΧΑ

Αὐρίλιος Μάρκος
κὲ Σίσινος Διάκων
κὲ Ἀλέξανδρος
ἀνεστήσαμεν τὸ-
ν τίτλον τοῦτον
Δουμετάῳ πρεσβυτέρῳ
μνήμης χά[ριν].

No. 216.

Konia. Copied by Dr. S. N. Diamantides.[1]

ΤΑCΙΟCΕΙΟΥΛΙΟ
ΠΑΤΡΙΚΙΟCΤШ
ΠΟΘΙΝΟΤΑΤШ
ΜΟΥΑΔΕΛΦШ
5 ΜΝΗCΙΘΕШΑ
ΝΕCΤΗCΑΤΗΝCΤΗΛΗΝ
ΤΑΥΤΗΝΜΝΗΜΗCΧΑΡΙΝ

---

[1] In line 7 NMNHMH are in ligature.

[Γ]ά[ε]ιος Εἰούλιο[ς]
Πατρίκιος τῷ
ποθινοτάτῳ
μου ἀδελφῷ
5 Μνησιθέῳ ἀ-
νέστησα τὴν στήλην
ταύτην μνήμης χάριν.

See No. 217.

No. 217.

Konia. Copied by Dr. S. N. Diamantides.[1]

ΓΑΕΙΟССΙΟΥΛΙΟС
ΠΑΤΡΙΚΙΟСΤΗΓΛΥ
ΚΥΤΑΤΗΜΟΥΘΙΑ
ΟΡΕСΤΙΝΕΝΚΡΑ
ΤΕΥСΑΜΕΝΗΑΝ
ΕСΤΗСΑΜΝΗ
ΜΗСΧΑΡΙΝ

Γάειος [Ε]ἰούλιος
Πατρίκιος τῇ γλυ-
κυτάτῃ μου θίᾳ
Ὀρεστί[δι] ἐνκρα-
τευσαμένῃ ἀν-
έστησα μνή-.
μης χάριν.

---

[1] Ligatures occur: line 3, ΗΜ; line 4, ΝΚ; line 5, ΑΜ.

No. 218.

Konia.   Copied by Dr. S. N. Diamantides: Ἐπὶ πέτρας εὑρισκομένης ἔν τινι ὀθωμανικῇ οἰκίᾳ· μηκ. 0.80 μ., πλ. 0.30 μ.

```
ΜΕΝΕΔΗΜ
Ο C Μ Ε Ν Ε Δ Η Μ
Ο Υ Κ Α Ι Α Ρ Ε C Κ
Ο Υ C Α Γ Υ Ν Η
Α Υ Τ Ο Υ Ζ ω Ν Τ Ε C
Μ Ν Η Μ Η C Χ Α
Ρ Ι Ν
```

Μενέδημ-
ος Μενεδήμ-
ου καὶ Ἀρέσκ-.
ουσα γυνὴ
αὐτοῦ ζῶντες
μνήμης χά-
ριν.

No. 219.

Konia.   Copied by Dr. S. N. Diamantides: Ἐπὶ πέτρας
ἐκτισμένης ἐντῷ τοῦ Μολλὰ Χιουγκιὰρ ἀναβρυτηρίῳ ἐν τῷ
Μεραμὶ· ὕψ. 0.80, πλ. 0.50.[1]

```
Ο Υ Α Λ Ε Ρ Ι Ο C
Κ Α Λ Λ Ι C Τ Ο Ν Γ Υ
Ν Ε Κ Ι Α Υ Τ Ο Υ Δ Ο
Μ Ν Ι Κ Ε Ε Θ Υ Γ Α Τ Ρ Ι
Μ Ε Ν Τ Ε Ι Ν Ι
```

---

[1] In line 4 ΚΕ are in ligature, an additional Ε being inserted besides.

Οὐαλέριος
Καλλίστο(υ) γυ-
νεκὶ αὐτοῦ Δό-
μνι κὲ⟨ε⟩ θυγατρὶ
Μεντείνι.

No. 220.

Konia. Copied by Dr. S. N. Diamantides: Ἐπὶ μαρμάρου
τετραγώνου ἐκτισμένου ἐν τῷ τοίχῳ τῆς αὐλαίας θύρας
ὀθωμανικοῦ τινος τεμένους κατὰ τὴν ὁδὸν Σετηρλὲρ, καὶ
φέροντος ἴχνη ἀναγλύφου· ὕψ. 0.80, πλ. 0.35.

K A Λ E Φ O Y Γ E I T Ω Γ Λ Y K Y T A T Ω M · · A N · K M
I O Y Λ I A N O C M N H M H C X A P I N

Καλεφούγει τῷ γλυκυτάτῳ μ[ου] ἀν[δρὶ]?· . . .
Ἰουλιανὸς μνήμης χάριν.

Καλεφούγει (-η)? is the name of the wife. Ἰουλιανός should be
Ἰουλιανῷ.

No. 221.

Konia. Copied by Dr. S. N. Diamantides: Ἐπὶ μαρμάρου
ἐστρωμένου ἐν τῇ ἀγορᾷ· ὕψ. 0.35, πλ. 0.55.

O Y Λ Π E I A K · O N H
K A I T Y C T E K N O I C M

Οὐλπεία Κ[λε]ον(ό)η?
καὶ τῦς τέκνοις μ[νήμης
χάριν].

No. 222.

Konia. Copied by Dr. S. N. Diamantides: Ἐπὶ τεμαχίου μαρμάρου ἐσπασμένου κατὰ τὴν βάσιν καὶ ἐκτισμένου ἐν τῷ τοίχῳ τῶν φυλακῶν· ὕψ. 0.24, πλ. 0.64.[1]

```
ΑΥΡΗΙΗΡΑΔΑΤΗCΕ
ΚΤΟΡΟCΖWΗ
ΝW
Ν
```

Αὐρή⟨ι⟩⟨λιος⟩. Ἡραδάτης? Ἕ-
κτορος Ζωῇ [γυναικὶ κτλ.]

. . . .

No. 223.

Konia. Copy of Dr. S. N. Diamantides, corrected by J. R. S. S.: Ἐπὶ τῆς προσόψεως στήλης ἑκατέρωθεν μόνον κυλινδρικῆς καὶ ὡς ὑποστήριγμα τοῦ Μουσάλλα Ταϊὴ χρησιμευούσης ἔμπροσθεν τοῦ τεμένους Σεραφεττήν· ὕψ. 0.70, πλ. 0.40.

```
ΔΟΜΝΟCΦΛΑ
ΒΙΟCΚΑΙ
ΛΙWΝΥΙΟCΑΥ
ΤΟΥΕΑΥΤΟΙC
ΖWCΙ
```

Δόμνος Φλ[ά-]
βιος καὶ [Πω-]
λίων υἱὸς αὐ-
τοῦ ἑαυτοῖς·
Ζῶσι.

---

[1] In line 1 ΤΗ are in ligature.

No. 224.

Konia. Copied by Dr. S. N. Diamantides: Ἐπὶ πέτρας λοξῶς ἐσπασμένης κατὰ τὴν ἀριστερὰν πλευρὰν καὶ εὑρισκομένης κατὰ τὸ Μεράμι· ὕψ. 1.0, πλ. 0.65, παχ. 0.25.

```
O C K Y A P Δ I O C K A I T P A Λ Λ E Y C
Δ H M·A Δ H C E K C A P Δ E I C·
T I · O C Δ E Δ O K            C
A N Δ P A C Π P O C K A P X H Δ O N
A N Δ P I A N T O C
```

. . . .

. . . . . . καὶ Τραλλεὺς

. . . .

Δημάδης ἐκ Σάρδε[ων?]
Τί[τ]ος? δὲ? Δοκ[ιμεὺς?] . . .
· ἄνδρας πρὸς Καρχηδόν[α]
ἀνδριάντος . . . .

No. 225.

Konia. Copied by Dr. S. N. Diamantides: Ἐπὶ λίθου παριστῶντος λέοντα, ὕψ. 1½, πλ. 1½, εὑρισκομένου ἐν τῇ οἰκίᾳ τοῦ Ἀρμενίου Παλθόγλου Ποβὸς καὶ ἐπὶ τῆς προσθίας ἐπιφανείας μεταξὺ τῶν ποδῶν τοῦ λέοντος ὑπὸ τὴν γαστέρα ἀμέσως φέροντος τὴν ἑξῆς ἐπιγραφήν.

```
O Y A Λ H C
K     M A N.N I C
Π A Π Π A Y O I
A N E C T H C A N T E P N A N
Λ A Λ Λ I A N Γ H N M H T E P A A Y T H C
```

Οὐάλης
κ[αὶ] Μάννις (= Μάννης)
Παππᾶ ὑοὶ
ἀνέστησαν Τέρναν
Λαλλίαν [τ]ὴν μητέρα αὐτῆς.

A St. Mannis is mentioned in an inscription of Iconium; see *Bulletin de Correspondance Hellénique*, 1883, p. 315.

No. 226.

Konia. Slab from recently demolished walls. Copy.

▨▨▨▨\IΛΠΑΝΧΑΡΙΟ(▨▨▨▨▨▨
▨▨▨▨)ΨΠΟΛΙΤΗCΟΨ▨▨▨▨▨▨
▨▨▨▨ΖΙCΕΜΝΟΤΑΤΗ ΪΪ▨▨▨▨
▨▨▨▨▨▨▨▨▨▨▨▨ΟΨΤΑΠΕΛΙ▨▨

. . . . Πανχάριο[ς] . . . .
['Αδριαν]ουπολίτης Οὐ[ειλίᾳ?]
[τῇ] σεμνοτάτῃ [γυναικὶ].

No. 227.

Konia. Copied by Dr. S. N. Diamantides: 'Επὶ στηλης τετραγώνου εὑρισκομένης ἔν τινι περιβόλῳ ἐν τῷ Μεράμι· ὔψ. 0.80 μ., πλ. 0.30 μ.

KOYTI·
ΔΗΜC
ΓΙCΤШ
ΕΥΧΗΝ

Κουτι?
Δ(ιὶ) Με-
γίστῳ
εὐχήν.

No. 228.

Konia. Copied by Dr. S. N. Diamantides; verified by J. R. S. S.: Ἐπὶ τεμαχίου λίθου πανταχόθεν κατεσπασμένου καὶ εὑρισκομένου ἐν τῇ οἰκίᾳ τοῦ κ. Λ. Κοσκινίδου· ὕψ. 0.20, πλ. 0.44, παχ. 0.25, ὕψος γραμμάτων 0.06, καὶ κεχρωματισμένην μὲ ἐρυθροῦν χρῶμα. *The color is now gone.*

[Ὁ δεῖνα τοῦ δεῖνος ἱερεὺς]
Διὸς μ̣ε̣[γίστου διὰ]
βίου καὶ ἀρ[χιερέα]
θε[ῶ]ν Σεβασ[τῶν].

```
ΔΙΟCΜΕ
ΒΙΟΥΚΑΙΑΡ
ΘΕΟΝΣΕΒΑΣ
```

No. 229.

Konia. Copied by Dr. S. N. Diamantides: Ἐπὶ λευκοτάτης μαρμαρίνου στήλης κωνοειδοῦς τὴν κορυφήν, τετραγώνου δὲ τὴν βάσιν, εὑρισκομένης πρὸ τῆς θύρας τῆς Ἐκκλησίας ἐν τῇ τοῦ Ἁγίου Χαρίτωνος Μανῇ περὶ τὴν μίαν ὥραν πέρ.που τοῦ Ἰκονίου· ὕψ. 0.50, μῆκ. 1.80, παχ. 0.25, ὕψ. γραμ. 0.09.

```
ΕΝΤΑΥΘΑΚΕΙΤΑΙΠΟΡΦΥΡΟΓΕΝΗΩΝΓΟΝΟCΗΙΧΑΗΛΑΗΗΛΑΗΡΑCΛΛΑΝΗΕΙΚΟΝΤΟΥ
ΠΑΝΕΥΓΕΝΤΑΒΑΙCΕΙΚΟΝ̃ΩΝΑΘΔΗΙΩΠΟΡΦΡΟΓΕΝΗΩΝΒΑCΙΛΕΩΥΚΥΡ8|ῶΚ·ḢΝΡ8ΤΡ
ḢΑΥΡο8 ωΗΗ · ΥΙΟCΔΕΒΤΑΤΕΙΝ8Ιῶ̃8Κ·ḢΝΡ8ΕΝΕΤΗΤΩΤ//////ΤΑΗΝΙΝΟCΗΒΡΙω
```

Ἐνταῦθα κεῖται Πορφυρογενητῶν γόνος Μιχαὴλ . . .

. . . . πανευγενεστάτου

. . . . Πορφυρογενητῶν βασιλέων

κύρου Ἰω(άννου) κ(αὶ) . . . . υἱὸς δὲ τοῦ

ταπείνου Ἰω(άννου) τοῦ . . . ἐν ἔτῃ . . . .

μηνὶ Νο[ε]μβρίῳ.

No. 230.

Konia. *Copied by Dr. S. N. Diamantides:* Ἐπὶ μαρμαρίνης πλακὸς εὑρισκομένης ἐν τῇ τοῦ Ἁγίου Χαρίτωνος Μονῇ· μηκ. 1.80, πλ. 0.23, παχ. 0.10, ὕψ. γραμ. 0.09.

ЄΚΩΗΗΘΗΟΔΟΥΛΟϹΤΟΥΘΥ·
ΑΒΡΑΑΗ[ΑΚΙϹ]ЄΤΟΥϹΛΟΗЄΗΥ

ἐκοιμήθη ὁ δοῦλος τοῦ θ(εο)ῦ
Ἀβραὰμ . . . . ἔτους . . .

No. 231.

Konia. *Copied by Dr. S. N. Diamantides:* Ἐπὶ πέτρας ἐρρι-μένης ἔμπροσθεν τῆς οἰκίας τοῦ Μετζητιέ· ὕψ. 0.80 μ., πλ. 0.40 μ.

†ΦΛΑΒΙΟϹΚΟΝШΝ
ΑΠΟΔΟΜЄΤΙΚШΝ
▓▓▓▓▓ΗδЄΤΡΙΑ
▓▓▓▓▓ЄΥΛΑΒЄϹ
ΚΟΝШΝ
▓▓▓▓ΑΝЄϹΤΗϹЄΝ
Μ.ΝΗΜΗϹΧΑΡΙΝ

Φλάβιος Κόνων
ἀπὸ δομε(σ)τίκων
. . . .
. . . . . εὐλαβὲς
. . . . . . Κόνων
. . . . ἀνέστησεν
·μνήμης χάριν.

No. 232.

Konia. Copied by Dr. S. N. Diamantides: Ἐπὶ μαρμάρου
εὑρισκομένου κατὰ τὴν ὁδὸν Τζελὰλ παρὰ τῇ αὐλαίᾳ θύρᾳ
τοῦ Ῥαφὲτ Τσελεπιάν.

ЄΝΤШΝΟШΤΟΥ
ΚΥΡΙΟΥΙΗϹΟΥ
ΧΡЄΙϹΤΟΥ▓▓▓
ϹΤΑΤΗΝΤΗ
ΜΝΗΜЄΙΔΟ
ΖΑϹΟΙΚΥ
ΡΙЄ▓▓▓

Ἐν τῷ ν[ε]ῳ΄ τοῦ
Κυρίου Ἰησοῦ
Χρειστοῦ . . . .
. . . . τῇ
μνήμει, δό-
ξα σοι Κύ-
ριε [εἰς ἀεί?].

If the conjecture in line 1 be correct, then the inscription dates
from the year 855 A.D.

No. 233.

Konia.   Copied by Dr. S. N. Diamantides: Ἐπὶ τεμαχίου
εὑρισκομένου κατὰ τὸ Μεράμι· ὕψ. 0.50, πλ. 0.35.

A N Δ P
A Λ E Ξ A
C Y M
Δ I O Γ E
M · A

. . . . . [τῷ]
ἀνδρ[ὶ] . . .
Ἀλεξά[νδρῳ]
συμ[βίῳ] . . .
Δίογεν . . . .

No. 234.

Konia.   Copied by Dr. S. N. Diamantides: Ἐπὶ πέτρας
ἑκατέροθεν ἐκ τῶν πλαγίων ἐσπασμένης καὶ εὑρισκομένης ἐν
τῷ Μεράμι ἔσωθεν τῆς αὐλαίας θύρας τοῦ χα" Σαταρεττὴν
οἰκίας καὶ χρησιμευούσης ὡς γέφυρα ἐπὶ ῥυακίου· ὕψ.
0.80 μ., πλ. 0.65 μ.

R S I B I E T M I
R I S V A E C V M
A I D I S A V T E M
V R S O L V S
V I O C T H P H T I

. . . . . sibi et Mi . . . . .
uxo]ri suae cum . . . . .'. .
. . . . autem . . . . .
. . . . ur solus . . . .
Φλάο]υιος Τηρητι[ανὸς.

No. 235.

Konia.   Copied by Dr. S. N. Diamantides: Ἐπὶ μαρμάρου
εὑρισκομένου ἔν τινι ὀθωμανικῇ οἰκίᾳ φέροντος ἐπὶ κεφαλῆς
σταυρὸν ἐγγεγραμμένον ἐντὸς κύκλου· ὕψ. 1.12, πλ. 0.30,
παχ. 0.46.[1]

```
   Δ Υ Ο Κ Α C Ι
   Γ Ν Η Τ Ο Ι
   Ι Ν Α Ν Α Ε
   Α C Ι ⲱ T▨
 5 Α Ι Θ Ε Κ Λ Η
   Μ Η Τ Η Ρ Η
   Δ Ε Π Α Τ Η Ρ
   Β Α Ρ Υ Π Ε
   Ν Θ Ο C Α
10 Μ Φ Ο Τ Ε Ρ Ο
   Ι C Α Ν Ε C Τ Η
   C Α C Τ Η Λ
   Η Ν Μ Ν Η Μ
   Ε Ι Ο Ν Ο Φ Ρ Α
15 Π Ε Λ Ο Ι Τ Ο
```

Δύο κασί·
γνητοι . . .

. . . . . .
. . . . . [κ-]
5 αὶ Θέκλη
μήτηρ ἡ
δὲ πατὴρ
βαρυπε
νθ[ὴ]ς ἀ

---

[1] In line 2 HT are in ligature.   Dr. D. regards lines 3 and 4 as doubtful.

10 μφοτέρο-
    ις ἀνέστη-
    σα στήλ-
    ην μνημ-
    εῖον ὄφρα
15 πέλοιτο.

No. 236.

Konia. *Copy of Dr. S. N. Diamantides, corrected by J. R. S. S.:*
Ἐπὶ λίθου ἐκτισμένου ἐν τῷ τοῦ Τερκιὰχ περιβόλῳ· ὕψ.
1.38, πλ. 0.46, παχ. 0.26.[1]

C H M A T I Ш Δ E
M I P O C Π P B C Y N A Λ O
H Π A C H Π I N Y T H C A O
K A Λ Λ I E K A I E P Γ O I C I
O Y T Ш C Ш C K A I Z Y N
M I H N Y K T I Θ A N O N
O I C T I T Λ O N E C T H C
A N T Ш N I O C K A I M A P I
M O Y C I K O I C Π E C C I I
O Y C Γ O N E A C T E I C A N T E
P A C E C T I Θ A N O N T Ш N

    Σῆμά τι ὧδε . . . .

    . . . .

    . . πάση(ς) πινυτῆς
    κάλλ[ει] καὶ ἔργοισι
    οὕτως ὡς καὶ ξὺν
    μίῃ νυκτὶ θανόν[τ-]

---

[1] Ligatures occur: line 3, HΠ, HΠ; line 6, HN; line 9, ΠE; line 10, NE,
TE, NT; line 11, NT.

οις? τίτλον ἔστησ[αν]
'Αντώνιος καὶ Μάρ[κος]
μούσικοι . . . . . [τ-]
οὓς γονέας τείσαντε[ς]
. . . . . ἐστι θανόντων.

No. 237.

Konia.   Copied by Dr. S. N. Diamantides: Ἐπὶ λίθου
ἐστρωμένου ἔν τινι βρύσει κατὰ τὴν ὀθωμανικὴν συνοικίαν·
ὕψ. 0.50, πλ. 0.30.[1]

```
Ε Α Ν Δ Ε Τ Ι Σ Τ Λ Σ
Τ Η Λ▨▨▨Ι Κ Η Σ
Θ Σ Ο▨▨▨Τ Α Κ▨▨
Μ Ν Ο Υ Σ▨Χ Ο Λ Ω Μ
Ι Ο▨Σ Ε Χ Ο Ι Τ Ο▨
Μ Ε Τ Α▨Ε Τ Ο Ν Ε
▨▨Α▨▨▨▨▨Α Ι Κ▨
Μ Ο Υ Θ Ν Η Τ Ο Ν
Μ Ο Υ Μ Η Δ Ι Ν Α Σ
▨▨Σ Χ Θ Η Ν Α
```

ἐὰν δέ τις τ[ὴν σ-]
τήλ[ην ἀδ]ικήσ[ῃ]
θ[ε]ο[ὺ]ς κα]τα[χθο]-
[νί]ους [κε]χολωμ[έ-]
[ν]ο[υ]ς ἔχοιτο
μετὰ [δ]ὲ τὸν ἐ-
. . . .
. . . θνητὸν
. . . μηδ[έ]να [ἐ-]
[ισα]χθῆναι?

---

[1] In line 4 MN are in ligature in the copy of Dr. D.

No. 238.

Ak Tcheshme. Copied by Dr. S. N. Diamantides : Ἐπὶ
πέτρας ἐσπασμένης ἔμπροσθεν οἰκίας ὀθωμανικῆς ἐν Ἀκ-
Τσεσμε· ὕψ. 0.50 μ., πλ. 0.50 μ.

```
▨▨▨€ I N
▨▨M O Y C I Λ O
▨▨N O C A N Δ € €
▨C T I C € Π I C B I A C
▨T A I Y Π O K € I C €
A I Φ I C K Ш Δ H N
P I A X € I Λ I A
. . . . . .
. . . . . ἐ]ὰν δὲ ἕ-
[τερό]ς τις ἐπισβιάσ-
[η]ται ὑποκείσε-
ται φίσκῳ δην-
άρια χείλια.
```

No. 239.

Konia. Copied by Dr. S. N. Diamantides : Ἐπὶ πέτρας
ἐσπασμένης κατὰ τὸ μέσον καὶ κειμένης κατὰ τὴν ὁδὸν
Μουσάλλα· ὕψ. 1.0, πλ. 0.60, παχ. 0.25.

```
▨▨▨▨▨▨
K A I € A Y T H
Z Ш 6 A A M N
H M H C
        X A P I N
. . . . . .
καὶ ἑαυτῇ
ζώσα⟨α⟩ μν-
ήμης
        χάριν.
```

No. 240.

Konia. Copied by Dr. S. N. Diamantides: Ἐπὶ τεμαχίου
πέτρας ἐκτισμένης ἐν ταῖς βαθμῖσιν ὀθωμανικῆς τινος
οἰκίας· ὕψ. 0.32, πλ. 0.31, παχ. 0.08.

▨▨Ε Ρ Ι Α Ν▨▨	[Οὐαλ]ερίαν
▨▨Ο Ε Κ)▨▨	. . . . . [ἀ-]
Ν Ε С Τ Η С Ε	νέστησε-
Ν С Τ Η Λ Λ▨	.ν στή̣λ⟨λ⟩[ην]
▨Ν Π Ε Ρ Τ▨	. . [ὑ]πὲρ . . .

No. 241.

Konia. Copied by Dr. S. N. Diamantides: Ἐπὶ τετραγώνου
στήλης ἀντιστρόφως κεχωσμένης ἔν τινι ὁδῷ κατὰ τὴν
ὀθωμανικὴν συνοικίαν· ὕψ. ἄνωθεν τοῦ ἐδάφους πλ. καὶ
παχ. 0.60.

▨▨Ι Ε Ρ Ε Υ С Τ Ο Υ▨Ρ Ο▨	ἱερεὺς τοῦ . . .
▨Α Ο Ε Ι Ο▨▨Β Ι Ο С	. . . . . . βίος
▨▨Ο▨Φ Ш Ν	. . . . . .

No. 242.

Konia. Copied by Dr. S. N. Diamantides: Ἐπὶ μαρμάρου
κυκλικοῦ ὀλίγον κατὰ τὴν μίαν πλευρὰν ἐσπασμένου εὑρισκομένου ἔν τινι ὀθωμανικῇ συνοικίᾳ, καὶ φέροντος ἐπὶ τῆς
κεφαλῆς ἐξαίσιον καλλιτεχνικόν τι ἐσπασμένον καὶ κεκολο-
βωμένον· ὕψ. 0.64, πλ. 0.64, παχ. 0.26.

Θ Ε Ο Ι С Κ Α Τ Α	.Θεοῖς κατα-
Χ Θ Ο Ν Ι Ο Ι С'	χθονίοις.

No. 243.

Konia. Copied by Dr. S. N. Diamantides: Τὸ αἴνιγμα τοῦ το αἰνίττεται ἴσως τὸν ὄνυχα (=ὄνυχ) γράμματος αἱρουμένου τὸ ο μένει καὶ γίνεται νὺξ, ὅτε δύεται ὁ ἥλιος·

ΑΥΧΕΝΟCΕΚΔΟΛΙΧΟΥΓΗΘΕΝΑΕΙΡΟΜΕΝΗCΦΑΙΡΗΔΩCΥΠΕΡ
ΑΥΛΟΝΕΕΙΔΟΜΑΙΗΝΔΕΜΑCΤΕΥΧCΗCΕΝΔΟΝΕΜΩΝΛΑΓΟΝΩΝ
ΜΗΤΡΟCΦΕΡωΠΑΤΕΡΑΟΥΜΕΡΟCΕΙΜΙΟΚΑΙΤΕΜΝΗΜΑΙCΙΔΗ
ΡΟCΓΡΑΜΜΑΤΟCΑΙΡΟΥΜΕΝΟΥΔΥΕΤΑΙΟΗΛΙΟC

The inscription seems to be suspicious, and for that reason I do not give the minuscules.

No. 244.

Konia. Copied by Dr. S. N. Diamantides: Ἐπὶ μαρμάρου ἐκτισμένου ἔσωθεν τοῦ τείχους τῆς τοῦ Χριστοῦ Μεταμορφώσεως Ἐκκλησίας παρὰ τὴν αὐλαίαν αὐτῆς θύραν· ὕψ. 0.90, πλ. 0.65.[1]

ΠΤΗΝΟΠΤΤΕΡΩΝΔΙΧΑΙΠΤΑCΘΑΙჳΚΕΝΙ
ΩCჳΔΕΓΡΟΝΜΘΑΙΤΙΟΥΓΕΝΕCΘΑΙ
ჳΠΡΟCΑΝΕΓΕΡCΙΝΘωΜᾶCΑΡS'ΑΚΕჳωΡ
ΤΟΥΠΙΚΛΗΝΟΚΑΛΟΥΜΕΝΟCΓSΤΜΤჳΖΑC
ΟΝΕΥΦΗΜΕΙჃΠᾶCΕΙCΙωΝΕΝΤΑῦΘᾶ
ΕΝΕΤΕΙ
ΑΨΛΓ

---

[1] Ligatures occur: line 1, ΗΝ, ΤΕ; line 2, ΝΕ; line 3, ΑΡ, ΑΚ; line 5, ΗΜ, ΤΩ.

No. 245.

Konia. Copied by Dr. S. N. Diamantides: Ἐπὶ γράμματα προυπάρξαντα. Τὸ αἴνιγμα τοῦτο αἰνίττεται ἴσως τὸν καπνὸν, τίκτοντα δάκρυα ἐκ τῶν ὀφθαλμῶν.

ΕΙΜΙΠΑΤΡΟΣΛΕΥΚΟΙΟ
ΜΕΛΑΝΤΕΚΟΣΑΣΤΕΡΟΣ
ΑΧΡΙΚΑΙΟΥΡΑΝΙωΝΙΠΤΑΜΕΝΟΣ
ΝΕΦΕωΝ
ΚΟΥΡΑΙΣΔΑΠΤΟΜΕΝΗΣΙΝΑΠΕΝ
ΘΕΑΔΑΚΡΥΑΤΙΚΤω

Εἰμὶ πατρὸς λευκοῖο μέλαν τέκος, ἀστέρος _ ∪∪
ἄχρι καὶ οὐρανίων ἱπτάμενος νεφέων
κούραις δαπτομένῃσιν ἀπενθέα δάκρυα τίκτω.

No. 246.

Konia. Copied by Dr. S. N. Diamantides: Ἐπὶ τετραγώνου στήλης κεχωσμένης κατὰ τὴν βάσιν παρὰ τῇ αὐλαίᾳ θύρᾳ Μολλὰ Χιουγκιὰρ· ὕψ. ἀπὸ τοῦ ἐδάφους 0.40; πλ. καὶ παχ. 0.35, ὕψ. γραμμάτων 0.055.

ΚΟΙΝΤΟΣΕ
▨ΟΥ▨ΗΝΟΣΜ
▨ΙΜΟΣ
ΝΕΜΕΣΕΙΣ
▨▨▨▨▨▨▨▨▨
▨▨▨▨▨▨▨▨▨

Κόϊντος . . .

No. 247.

*Konia. Panel on a slab from the recently demolished walls.
It is broken down the centre, the right half being gone.*

```
✸ΤΟΝΨΥΧ
ΤΙΓΕΝΝΕ
ΒΟΥΡΙΚΕ
ΤΟΔΕΦΥΓ
ΟΤΕΧΡΙϹ
ϹΟΝΟΙΚΗ
```

No. 248.

Konia. Copied by Dr. S. N. Diamantides: Ἐπὶ μαρμαρίνου
στήλης ἀντιστρόφως κεχωσμένης παρά τινι βρύσει κατὰ
τὴν ὀθωμανικὴν συνοικίαν.

```
ΤΟΥ    ΟΝΑΠΟΝΕΡΓ
ΒΟ  Ο  ΕΧΕΙΝ
ΔΕΕΞΟΥϹΙΑ
ΝΤΗΝΑΒΟΥ
ΑΟ    ΙΚΙΔΕ
ΠΡΟϹΟΑ
ΟΝ    ΤΛΥ
[Τ]ΕΚΝΑ
```

No. 249.

Konia. Copied by Dr. S. N. Diamantides: Ἐπὶ τεμαχίου
πέτρας ἐστρωμένης ἐπὶ τοῦ λιθοστρώτου τῆς ἀγορᾶς· ὕψ.
0.60, πλ. 0.40.

```
ΜΕΙΡΟΙΜΟΥ
ΝΑΓΥΜΑΙΚ  ΙΜ
        ΛΙΚΙΝ
```

No. 250.

Konia.   Copied by Dr. S. N. Diamantides: Ἐπὶ πέτρας καθέτως κατὰ τὸ μέσον ἐσπασμένης καὶ εὑρισκομένης ἔν τινι ὁδῷ κατὰ τὸ Τοουκιοῦρ Τσεσμέ· ὕψ. 0.80, πλ. 0.26, παχ. 0.26.

```
        Θ Ο Ο Α
        Η C Μ
        Α Λ Є
        Υ Δ Ι Ο C
        Ν Τ Ш
        Ο Υ·Є Ι
        Τ Ρ Є C
```

No. 251.

Konia.   Copied by Dr. S. N. Diamantides: Ἐπὶ τῆς πλευρᾶς τετραγώνου μαρμάρου εὑρισκομένου ἔν τινι συνοικίᾳ ὀθωμανικῇ παρὰ τῷ στρατῶνι (Κήᾶλα).

```
O E T ῥϛ M ῥϛ O P P I O
```

No. 252.

Adalia.   Copied by Dr. S. N. Diamantides, who only remarks that it is not far from Adalia.

```
  Α Υ Τ Ο Κ Ρ Α Τ Ο Ρ Ι Κ Α Ι Σ Α Ρ Ι Θ Ε
  Ο Υ Α Δ Ρ Ι Α Ν Ο Υ Υ Ι ϾϿ Θ Ε Ο Υ Τ Ρ Α Ι Α
  Ν Ο Υ Π Α Ρ Θ Ι Κ Ο Υ Υ Ι ϾϿ Ν ϾϿ Θ Ε Ο Υ
  Ν Ε Ρ Ο Υ Σ Ε Γ Γ Ο Ν ϾϿ Τ Ι Τ · ϾϿ Α Ι Ν ϾϿ
5 Α Δ Ρ Ι Α Ν ϾϿ Α Ν Τ Ω Ν Ε Ι Ν Ш Σ Ε Β Α Σ
  Τ Ш Ε Υ Σ Ε Β Ε Ι Α Ρ Χ Ι Ε Ρ Ε·Ι Μ Ε Γ Ι
```

ΣΤΩΔΗΜΑΡΧΙΚΗΣΕΞΟΥΣΙΑΣ
ΤΟΥΠΑΤΩΤΟΔΠΑΤΡΙΠΑΤΡΙ
ΔΟΣΚΑΙΘΕΟΙΣΣΕΒΑΣΤΟΙΣΚΑΙ
10 ΤΟΙΣΠΑΤΡΩΟΙΣΘΕΟΙΣΚΑΙΤΗ
ΓΛΥΚΥΤΑΤΗΠΑΤΡΙΔΗΤΗΠΑ
ΑΡΑΩΝΙΠΟΕΙΤΗΜΗΤΡΟΠΟΛΙΟΥ
ΛΥΚΙΩΝΕΘΝΟΥΣΟΥΕΙΛΙΑΚΟ.
ΟΥΕΙΛΙΟΥΤΙΤΙΟΝΟΥΘΥΓΑΤΗΡ
15 ΚΑΙΚΑΘΙΕΡΩΣΕΝΤΟΤΕΠΡΟΣΚΗ
ΝΙΟΝ.
ΟΚΑΤΕΣΚΕΥΑΣΕΝΕΚΘΕΜΕΛΙΩΝ
ΟΠΑΤΗΡΑΥΤΗΣΚΟΟΥΕΙΛΙΟΣ
ΤΙΤΙΩΝΑΣΚΑΙΤΟΝΕΝΑΥΤΩ
20 ΚΟΣΜΟΝΚΑΙΤΑΙΕΡΙΑ▨▨▨ΟΚΑΙ
ΤΗΝΤΩΝΑΝΔΡΙΑΝΤΩΝΚΑΙΑΓΑ
ΛΜΑΤΩΝΑΝΑΣΤΑΣΙΝΚΑΙΤΗΝ
ΤΟΥΛΑΓΕΙΟΥΚΑΤΑΣΚΕΥΗΝ
ΚΑΙΠΛΑΚΩΣΙ▨▨ΑΕΠΟΙΗΣΕΝ
25 ΑΥΤΗΤΟΔΕ▨▨▨Δ.ΕΚΑΤΟΝ▨▨▨
ΟΥΔΙΑΖΩ▨▨ΑΤΡΟΣΒΑΘΡΟΝ
ΚΑΙΤΑΒΗΛΑΤΟΥΘΕΑΤΡΟΥ
ΚΑΤΑΣΚΕΥΑΣΘΕΝΤΑΥΠΟΤΕΤΟΥ
ΠΑΤΡΟΣΑΥΤΗΣΚΑΙΥΠΑΥΤΗΣ
30 ΠΡΟΑΝΕΤΕΘΗΚΑΙΠΑΡΕΔΟΘΗ
ΚΑΤΑΤΟΥΠΟΤΗΣΠΡΟΤΙΣΤΗΣ
ΒΟΥΛΗΣΨΗΦΙΣΜΕΝΟ.

Αὐτοκράτορι Καίσαρι, θεοῦ
Ἀδριανοῦ υἱῷ, θεοῦ Τραϊα-
νοῦ Παρθικοῦ υἱωνῷ, θεοῦ
Νέρου(α) ἐγγόνῳ, Τίτῳ Αἰ[λί]ῳ
5 Ἀδριανῷ Ἀντωνείνῳ Σεβαστῷ Εὐσεβεῖ, ἀρχιερεῖ μεγί-

στῳ, δημαρχικῆς ἐξουσίας
τὸ (ή), ὑπάτῳ τὸ δ΄, πατρὶ πατρίδος καὶ θεοῖς Σεβαστοῖς καὶ
10   τοῖς πατρῴοις θεοῖς καὶ τῇ
γλυκυτάτῃ πατρίδ(ι) τῇ Πα[τ-]
αράων πό[λ]ει τῇ μητροπόλ[ει] (τ)οῦ [doubtful]
Λυκίων ἔθνους Οὐειλία Κο(ίντου)
Οὐειλίου Τιτι[ώ]νου θυγάτηρ
[A line has been omitted by the copier.]
15   καὶ καθιέρωσεν τό τε προσκή-
νιον.
ὃ κατεσκεύασεν ἐκ θεμελίων
ὁ πατὴρ αὐτῆς Κό(ίντος) Οὐείλιος
Τιτιώνας καὶ τὸν ἐν αὐτῷ
20   κόσμον καὶ τὰ ἱερ(ε)ῖα . . . . καὶ
τὴν τῶν ἀνδριάντων καὶ ἀγα-
λμάτων ἀνάστασιν καὶ τὴν
τοῦ λαγείου? κατασκευὴν
καὶ πλάκωσι[ν]? . . . . ἐποίησεν
25   αὐτὴ το . . . . δέκατον
· . . . . βάθρον
καὶ τὰ βῆλα τοῦ θεάτρου
κατασκευασθέντα ὑπό τε τοῦ
πατρὸς αὐτῆς καὶ ὑπ᾽ αὐτῆς
30   προανετέθη καὶ παρεδόθη
κατὰ τὸ ὑπὸ τῆς πρ[ω]τίστης
βουλῆς ψηφισμένο[ν].

We spent two days in Konia copying inscriptions and taking photographs of the Seldjuk city. The people of this eastern country seem to have had little interest in the affairs of this world, and spent their surplus energy in preparing tombs and epitaphs for themselves ; witness the above inscriptions. When Leake passed through Konia, the walls of the city were full of inscriptions, which he had no time to copy. After the destruction of Konia by the Egyptians, under Ibrahim Pasha in 1833, these walls were used as quarries for the modern city of Konia. The inscriptions seen by Leake have all perished in this way before an epigraphist was found to copy them. But many inscriptions are no doubt still in the walls that remain, with the inscribed side hidden from view. Part of the wall had been thrown down only a short time previous to our visit, and I copied several inscriptions that had been brought to light in this way. These walls, though most probably of Seldjuk origin, were built in the common Greek fashion (Thuc. I. 93) ; that is, two walls were built at a fixed distance apart, and the space between them was filled with earth and stone débris. At Konia the filling consisted mostly of simple clay or mud, which took faithful impressions of the stones composing the outer shell of the wall, so that one may now see therein neat reliefs of inscriptions, Phrygian doors, and architectural fragments. The ruins of the buildings erected by the early Seldjuk Sultans of Konia speak in elegant terms of former splendor.

The Governor of the Vilayet of Konia, Saïd Pasha, who studied in England and speaks English fluently, showed us kind attentions in more ways than one. He is collecting the most important antiquities of the district, as they come to light, for the Imperial Museum in Constantinople, and the collection is not without interest. Among other things may be mentioned a frieze in very high relief. Unfortunately we were unable to get photographs of the collection.

July 5. Konia to Obruk, 9 h. 47 m. The road from Konia to Ak Serai, the ancient Archelaïs, crosses the desert region. The plain, up to the pass in Boz Dagh, is absolutely level, and the thirsty traveller is mocked on all sides by the *Fata Morgana*, promising water near at hand ; but the promised water recedes continually, and finally turns out to be nothing but a deceptive mirage. We did not think it necessary to water our horses at Zeïvedjik, and consequently they had to make the whole long journey to Obruk thirsting.

No. 253.

Sindjerli Khan.   Round column.   Copy.

```
KANAI C░░░░░░░░░░
░░CⓈNΓEPAIOIKA░
MANIONΠAΓIK░░
TONEAYTWNEY░
THNTWT░EI░
THPIWKAITW░
ETEIMHΣAN░
```

. . . .

. . . . . .

μάνιον Πασικ[λέα]
τὸν ἑαυτῶν εὐ[εργέ-]
5 την . . . .
τηρίῳ καὶ τῷ . . . .
ἐτείμησαν . . . . . .

This is a ruined Khan with no water.

No. 254.

Doksan Dokus Merdimenli Kuyu, east of Sindjerli Khan.
The second step of the well.   Copy.

```
Γ·AΠΠΩNIOΣKPIΣΠOΣↃ░░░
EIKONIΩYKAIAIAIΛIAΔAΔAHΓ░
AIΛIΩIOYΛIANΩTEIMΩΘEΩYYIΩT░
TΩYΓYNAIKIMNHMHΣXAPIN░
EΞEΣTAIEIΣKOMIΣΘHNAIΣOP░
MIOYKAIΔIΔΩΓYNAIKATΩ░
```

Γ. Ἀππώνιος Κρίσπος [Δ . . . . .]
Εἰκονίων καὶ Αἰλία Δάδα ἡ γ[υνὴ αὐτοῦ]
Αἰλίῳ Ἰουλιανῷ Τειμωθέων υἱῷ τ . . . .
των γυναικὶ μνήμης χάριν· [οὐδενὶ δὲ]
ἔξεσται εἰσκομισθῆναι σο[ρὸν] . . . . .
μίου καὶ Διδῶ γυναικὶ τω . . . .

The well bears the name : "*well with the ninety-nine steps.*" The steps leading down to the water are still in situ; the water is brackish. At the ruined and deserted Dibidelik Khan there is a great well, both with a vertical well-shaft, and with a tunnel leading down to the water at an angle of about 30°. The water cannot be drunk by man or beast. This point is the limit of the waterless and hence desert plain of Konia.

At Obruk there is a little lake, the surface of which is about ninety feet below the surrounding country. The villagers use the water of this lake for household purposes. We were told that the water is drinkable at all seasons of the year, except for two weeks in December, when it is in a state of violent ebullition. When this season approaches, the villagers lay in a supply of water sufficient to last until the lake has resumed its wonted calm. How true this may be, or what causes the ebullition of the water, I am not prepared to say.

July 6. Obruk to Sultan Khan, 7 h. 31 m. The country is not a level plain, but is gently rolling ground. The land would everywhere be productive if it could only be irrigated. Unfortunately, water can be had only at intervals, for instance, at the villages Orta-kuyu, Bakharakh, Erdodu. These villages raise crops that can do with the winter and spring rains, such as wheat and barley, but they rely mainly on their herds for subsistence.

Sultan Khan is the grandest and most beautiful of all the remains of Seldjuk splendor seen by us in Asia Minor. We spent one day in its welcome shade, during which time numerous photographs were taken, and the huge building was roughly measured. One of the Arabic inscriptions says that it was built in 1277 A.D. A very large spring rises near Sultan Khan, and the land yields abundant harvests wherever it can be properly irrigated. Indeed, this is true almost everywhere in Asia Minor.

July 8. Sultan Khan to Ak Serai, 7 h. 53 m. We pass the ruins
of a Seldjuk Khan in 4 h. 15 m. from Sultan Khan. Ak Serai is a
sleepy uninteresting town, with but few traces of the Graeco-Roman
civilization ; but the foot-prints of the Seldjuks are abundant.

Henceforth the reader may consult the map of Southern Cappadocia, which accompanies this volume.

Nos. 255–256.

Ak Serai (Archelaïs).   In a house.   Copy.

A.	B.
A T A K I N	A Γ I O C
A C T I K I	
C ˢ A Π O	
T O Y T O Y	† X̄ Ē C Y Γ
5 M A P I A H	X ω P H C O N
T A T H †	T A Π Λ H M̄ E
A N A Θ E M A	5 Λ H M A T A T ω̄
I T O K Y M H	K A T A K I M E
T O Y T O †	N ω N †
	O Θ̄C̄ C Π Λ
	A Γ X N I C
	10 A Γ▓▓▓▓▓
	O▓▓▓▓▓

B.

† Χ(ύρι)ε συγ-
χώρησον
τὰ πλημμε-
5 λήματα τῶν
κατακιμέ-
νων.
Ὁ θ(εὸ)ς σπλ-
αγχνίσ(εται?)
10 Ἄγ[ιον?].

No. 257.

Ak Serai. Copy.

P R O V

July 9. Ak Serai to Selme, 4 h. 4 m. At Selme we found numerous dwellings cut in the rock, similar to those described by the early travellers at Soghanlü Dere and Udjessar. In fact, we found such rock-cut dwellings wherever the soft volcanic tufa appears (Hamilton, *Researches*, I. p. 97). Selme is situated in a deep gorge, through which the Irmak flows, and in which, in fact, it has its source. The cliff to the east rises perpendicularly to a height of from four to five hundred feet; at its base there is a maze of sharp natural cones, similar to those in the region around Udjessar. Most of these cones have been excavated for human dwellings, often with several stories. These excavations are used as dwellings now, as in ancient times. The whole cliff is honey-combed into dwellings, chambers, chapels, passages, and tombs; story rising upon story. Here and there may be seen small temple façades on the exterior of the cliffs, especially at Ikhlara. These façades resemble those in the region of the tomb of Midas. People still live and die in these rock-cut dwellings, at least two hundred feet high on the cliff. There is no earthly reason why they should live there, as the country is safe and land abundant; but they do not seem to object to the dark winding stairs and passages.

Across the Irmak, five or ten minutes south of Selme, is the village of Ikhlara, the cliff behind which is also similarly honey-combed into dwellings. Conspicuous on the exterior of the cliff are temple façades, which were doubtless intended for tombs. A short distance east of Ikhlara the Irmak gushes out at the foot of the cliffs, a full-grown river at its source.

July 10. Selme to Kuyulu Tatlar, 4 h. 49 m. Leaving Selme we immediately ascend the bluff, and henceforward traverse an open country. The old map of this region, founded on von Moltke's flying ride, and the new sketch founded on the travels of Vrontchenko and Tchihatcheff, are all wrong. The map accompanying this volume will be found to be more accurate, it is hoped.

July 11. Kuyulu Tatlar to Ortakieui, 5 h. 34 m. Kuyulu Tatlar is so called from twelve or fifteen wells, all in a cluster. From them the village is supplied with water. This region, though blank on the old map, has numerous villages. It may be noted that the Tada Su of the old map does not exist, at least not in the plain of Kuyulu Tatlar and Malagob ; and, furthermore, the drain-water from this district must run south, and not north, as on the old map. Malagob is a large and flourishing village, whose inhabitants are in the main Greek-speaking Greeks. The Greeks are numerous all through the western part of Cappadocia. As a general rule they cling to their language with great tenacity, a fact worthy of notice, inasmuch as the Greeks in other parts of Asia Minor speak only Turkish. Their dialect has been treated by Καρολίδης in the Μουσεῖον καὶ Βιβλιοθήκη τῆς Εὐαγγελικῆς Σχολῆς, published in Smyrna. Instances of Greek-speaking towns or villages are Nigde, Gelvere, Malagob, and Orta-kieui, in what is commonly but wrongly called Soghanlü Dere. Leaving Malagob we shortly ascend a hill, which turns out to be the rim or bluff of an elevated plateau, extending east as far as Develü Kara Hissar. Soghanlü Dere, Ortakieui Dere, and what other Deres there may be, are mere breaks in this plateau, and the top of the bluffs of all the Deres correspond with the general level of the surrounding plateau. The descent down into Ortakieui Dere is made by an artificial road that has been excavated out of the soft tufa.

July 12. Ortakieui to Develü Kara Hissar. We got lost between Ortakieui and Soghanlü Dere, and the exact time cannot be given. The time between Balak and Develü Kara Hissar is 1 h. 50 m.

The wonders of Soghanlü Dere have been described by Hamilton. The rock-cut dwellings are more numerous, but of the same character as those at Selme and Ikhlara ; only at Soghanlü Dere there are no temple façades to be seen. At Bashkieui the Ortakieui Dere is about one hundred yards wide ; but the width increases steadily, reaching a width of from five to seven hundred yards at the point where Soghanlü Dere branches off laterally from it. While the surrounding plateau is a barren waste, the soil in the Deres is exceedingly fertile, delighting the eye with its luxuriant gardens. This is especially the case at Ortakieui. Whether these rock-cut habitations date originally from an earlier epoch or not, it is at all events certain that they were used by the early Christians. But such habitation goes back to a period so remote that the Christian Greeks of Ortakieui have no traditions concerning it. Chapels are numerous, in some of which may still be seen pictures of Byzantine Saints, with inscriptions just like those common in orthodox churches of to-day. Among the Saints depicted are Σέργιος, Βάχος, Μερκούριος (see Nos. 258 and 261). In the floor of the chapels graves were cut, in some of which we found human skeletons. Indeed, such tombs are frequent in the dwellings themselves, so that, as Hamilton remarks, the people lived in the same room with their pigeons and their dead. Innumerable pigeons live in the rock-cut dwellings both of Soghanlü Dere and Udjessar. At the latter place the villagers pay great attention to them and use them for food.

No. 258.

Soghanlü Dere. In a niche in a chapel. Copy.

```
A    C    O    B
     E    A    A
     P    Γ    X
     Γ    I    O
     I    O    C
     O    C
     C
```

῎Α(γιος) Σέργιος.

῾Ο ῎Αγιος Βάχος.

No. 259.

Soghanlü Dere. In a niche as above. Copy.

```
Δ Ε Ι C Ι C Τ Η Δ Ȣ
Λ Ι C Τ Ȣ Ⓨ⊖ Υ Ε Υ
Δ Ο Κ Ι Α C
```

Δέισις τῆ(ς) δού-
λις τοῦ θ(εο)ῦ Εὐ-
δοκίας.

No. 260.

Soghanlü Dere.   In a niche as above.   Copy.

Δ Ε Ι C Ι C T ४ Δ ४
Λ ४ T ४ ⊖ ४ N Y N Φ O
N O C M O N A X ४

Δέισις τοῦ δου-
λου τοῦ θ(εο)ῦ Νύνφο-
νος μοναχοῦ.

No. 261.

Soghanlü Dere.   In a niche as above.   Copy.

(A) M Ε P K ४ P Ι ४

'Α(γίου) Μερκουρίου.

Zengibar Kalesi is situated about half an hour west of Develü Kara Hissar.   It is a lofty rock with two peaks, one of which is consider-ably higher than the other.   In the saddle between the two peaks nestles Kalekieui.   There can scarcely be a doubt but that the higher peak of Zengibar Kalesi is Nova, the proud rock where Eumenes and his little band defied the whole army of Antigonus for nearly a year.

July 14.   Develü Kara Hissar to Indjesu, 4 h. 53 m.   We traverse the new road.   The country is desolate.   We suffered much from the intense cold, in spite of the fact that to-day is July 14th.

July 15.   Indjesu, *via* Kaisariye, to Talas, 6 h. 29 m.   We cross the southern end of the great Sazlik, or *place of the bulrushes*, fixing its coast line.

July 16.   Talas to Ispile, 1 h. 20 m.   Parting with regret from our kind friends, the American missionaries of Caesarea, we hurried on to the region east of the Antitaurus.

July 17. Ispile to Yokara Suvergen, on the eastern bank of the Zamantia Tchai, 6 h. 59 m. The country northwest of Tomarza is uninteresting and barren, except in the Deres, in which all the villages are situated. Tomarza is a large Armenian town, with considerable traces of ancient remains, most probably Armenian. At Sheikh Barakh we reached the Zamantia Tchai, which is spanned by a bridge at this point. We forded the river opposite Ashagha Suvergen; it is deep and rapid.

July 18. Yokara Suvergen to Ak Puñar, 7 h. 37 m. We crossed the Antitaurus by the precipitous pass between Dede Dagh and Bei Dagh. The region east of the Antitaurus was hitherto unknown; we found it fertile and well-populated. The inhabitants are Avshars and Circassians. These Circassians are refugees from Circassian Russia, and were assigned homes here by the Turkish government. Here, as everywhere, they have the reputation of being great cut-throats and robbers, but we were treated with the most distinguished consideration and kindness by them. Afterwards we visited numerous Circassian villages, and we have the same good report to give of them everywhere. I am told, however, that this was due to the fact that my chief servant was a Circassian. It is always pleasant to enter a Circassian village, for everywhere one sees order, thrift, and cleanliness, a refreshing treat after a prolonged sojourn in the wretched hovels of the Turkish peasantry.

July 19. Ak Puñar to Shahr, 3 h. 20 m. In 2 h. 25 m. the plain closes in to a wild and rugged gorge. The river Seihûn, the ancient Sarus, has cut its way through the mountains in many curves, each curve corresponding to a projecting spur of the mountain. Shahr, the ancient Comana, is the only place marked on Tchihatcheff's map south of Olakaya. The great goddess Ma is no longer worshipped at Comana; but, to our great astonishment, we found a Protestant church there, composed of the converts of the American missionaries. We spent a day here copying inscriptions. We also ascended Külek Dagh, on the summit of which we found a large and impregnable fortress of great antiquity, antedating the Roman conquest, most probably.

No. 262.

Shahr (*Comana*).   Bulletin de Correspondance Hellénique,
1883, *p.* 131. *Copy.*

```
ΙΕΡΟΠΟΛΕΙΤШΝ
    ΗΒΟΥΛΗ
  ΚΑΙΟΔΗΜΟC
   ΘΕΜΙCΤΟΚΛΕΑ
5  ΑΛΕΞΑΝΔΡΟΥ
  ΤΟΝΦΙΛΟΠΑΤΡΙΝ
```

Ἱεροπολειτῶν
ἡ βουλὴ
καὶ ὁ δῆμος
Θεμιστοκλέα
5 Ἀλεξάνδρου
τὸν φιλόπατριν.

Var. Lect.

Line 1.   The *Bulletin* reads : ΕΡΟΠΟΛ · · · · · ·

No. 263.

Shahr. *In the church.* Bulletin de Correspondance Hellénique, 1883, *p.* 127 ; Journal of Philology, *XI. p.* 147.
Copy.

```
                    ΜΗΙ
                   ΙΤΕΙ
              ΛΜΗΝΙΑΖΗΜ
          ΑΤΗΞΟΙΚΗΦΟΡΟΥΘΕ
        ΞΤΡΑΤΗΓΟΝΚΑΤΑΟΝΙΑ
      ΓΗΞΑΜΕΝΟΝΑΥΤШΝΕΠΙΕΙΚШ
     ΚΑΙΕΥΕΡΓΕΤΙΚШΞ         ·
```

. . . .
[ἱερέ]α τῆς (Ν)ικηφόρου Θε-
5 [ᾶς καὶ] στρατηγὸν Καταονία-
[s, ἡ]γησάμενον αὐτῶν ἐπιεικῶ-
[s] καὶ εὐεργετικῶς.

Var. Lect.

Line 3 init. The *Journal* and *Bulletin* omit ⫴.
Line 4 fin.   "   "   "   "   read L.
Line 6 init.   "   "   "   "   omit Γ, and read O at the
end.

No. 264.

Shahr.   Bulletin de Correspondance Hellénique, 1883,
 p. 131; Journal of Philology, *XI. p.* 149. *Copy.*

AYTOKPATOPA
KAICAPAMAYP
ΠI⧄⧄⧄ONEYCEB
EYTYXCEBMEΓ

Αὐτοκράτορα
Καίσαρα Μ. Αὐρ.
Π[ρόβ]ον Εὐσεβ(ῆ)
Εὐτυχ(ῆ) Σεβ(αστὸν) Μέγ(ιστον).

Var. Lect.

Line 3.   The *Journal* and *Bulletin* omit ☐.

No. 265.

Shahr.   Bulletin de Correspondance Hellénique, 1883,
 p. 135; Journal of Philology, *XI. p.* 160. *Copy.*

ΦΛ·ΑCΙΑΤΙΚΟC
ΚΑΙΙ☐ΥΛΙΑΑΘΗ
ΝΑΙCΠΑΠΟΥΦΛ
ΝΥCΗΤΗΓΛΥΚΥ

5 ΤΑΤΗΚΕΜ☐ΝΗ
ΑⅭΥΝΚΡΙΤШΘΥ
ΓΑΤΡΙ𝚷Ρ☐ΜΟΙ
Ρ𝖶

Φλ. ’Ασιατικὸς
καὶ ’Ιουλία ’Αθη-
ναῖς Πάπου Φλ.
Νύσῃ τῇ γλυκυ-
5 τάτῃ κὲ μόνῃ
ἀσυνκρίτῳ θυ-
γατρὶ προμοί-
ρῳ.

Var. Lect.

Line 6 fin. The *Journal* and *Bulletin* read ΟΥ.
Line 7 init. "   "   "   "   omit Γ, and the Ρ after the 𝚷.

No. 266.

Shahr. Bulletin de Correspondance Hellénique, 1883, *p.* 138; Journal of Philology, *XI. p.* 148. *Copy.*

A.

ΜΝΗΜ·ΑⅭΚΛΗ𝚷ΙΑΔ▨
𝚷ΥΛΑΔΟΥΤΟΔΕ
ΤΕΥΞΕΝΑΡΕΙШΝ

𝚷ΡШΤΟⅭΚΑΙΦΙΛΙ▨
5 ΚΑΙΓΕΝΕΙΕΝΓΥΤ Ι▨

B.

ΛΕΥΤΕΡΟⅭΑΥΘ·ΕΤΑΡШΝ
𝚷ΡΟΦΕΡШΝ
ΑⅭΚΛΗ𝚷ΙΟΔШΡΟⅭ

▨ΟΙΚΕΙΟⅭΦΙΛΙΗΝ
ΙΔΕ𝚷ΑΡШΝΥΜΙΗΝ

η

C.

ΦΑΙΔΡΟϹΔ░░ΥΤ
ΕΠΙΤΟΙϹΙΤΡ░░ΤΟϹ
ΦΙΛΙΗΔΑΡΑ·ΠΡѠΤΟϹ

ΔΕΙΜΑΤ·ΑΕΙΜΝΗϹΤΟΝ
15 ϹΗΜΑΦΙΛѠΕΤΑΡѠ

D.

ΤΕΤΡΑΤΟϹΑΥΜΕΜΦΙϹ
ΟΥΤΟΙΤΑΦΟΝ
ΕΞΕΤΕΛΕϹϹΑΝ

ΤΕϹϹΑΡΕϹΕΚΠΟΛΛѠΝ
20 ΜΝΗΜΟΝΕϹΕΥϹΕΒΙΗϹ

Μνῆμ' Ἀσκληπιάδ[η] Πυλάδου τόδε τεῦξεν Ἀρείων
Πρῶτος καὶ φιλί[η] καὶ γένει ἐνγύτ[ατος].
Δεύτερος αὖθ' ἑτάρων προφέρων Ἀσκληπιόδωρος
Οἰκεῖος φιλίην [ἠ]δὲ παρωνυμίην ·
Φαῖδρος δ' [α]ὖτ' ἐπὶ τοῖσι τρίτος, φιλίῃ δ' ἄρα πρῶτος,
Δείματ' ἀείμνηστον σῆμα φίλῳ ἑτάρῳ.
Τέτρατος αὖ Μέμφις · Οὗτοι τάφον ἐξετέλεσσαν,
Τέσσαρες ἐκ πολλῶν μνήμονες εὐσεβίης.

Var. Lect.

Line 1. The *Journal* and *Bulletin* omit the point, and do not indicate a break at the end.

Line 5. The *Journal* and *Bulletin* read ΚΑ · , and omit /░░ at the end.

Line 6. The *Journal* and *Bulletin* omit the point.

Line 11. The *Journal* and *Bulletin* read ΔΑΥΤ, and do not indicate a break in line 12.

Line 14. The *Journal* and *Bulletin* omit the point.

No. 267.

Shahr.   Stele in the mill.   Length, 0.72 *m.; width,* 0.30 *m.
Copy.*[1]

```
I A Ϲ Ш N A Π I
Ш N ☐ Ϲ A Θ H
N A I Δ I Γ Λ Υ
K Υ T A T H Ϻ ☐ Υ
Γ Υ N E K I K A I
E A Υ T Ш Ϻ N H
Ϻ H Ϲ X A P I N
```

’Ιάσων ’Απί-
ωνος ’Αθη-
ναΐδι γλυ-
κυτάτῃ μου
γυνεκὶ καὶ
ἑαυτῷ μνή-
μης χάριν.

No. 268.

Shahr.   In a house.   Length, 0.45 *m.; width,* 0.23 *m.   Copy.*[2]

```
A Υ P H Δ
I Ϲ T O C H
Δ I H T H Γ Λ
Υ K Υ T A T H
Θ Υ Γ A T P I
```

---

[1] Ligatures occur: line 1, ШN; line 2, ШN☐Ϲ; line 4, TH; line 5, NE;
line 6, TШ, NH; line 7, HϹ.
[2] Ligatures occur: line 3, TH; line 4, TH.

Αὐρ. Ἡδ-
ιστος Ἡ-
δίῃ τῇ γλ-
υκυτάτῃ
θυγατρί.

The ruins of Comana are by no means extensive. Chief among them are the temple, the ruins of the theatre, and a highly ornamental portal.

Comana was once so rich in temples and brilliant edifices that it bore the name of the "Golden." Even in the time of the first crusaders it was *pulcherrima* and *opima*. For the line of march of the first crusade, see Ritter, *Klein-Asien*, II. p. 265–272.

July 21. Shahr to Hadjin, 8 h. 12 m. We turn our faces toward the south. This day was rich in topographical results, and the map of Tchihatcheff was found to be wrong in almost every particular; see the map accompanying this volume. Hadjin is on the right, not on the left side of the Seihûn, as is the case on Tchihatcheff's map. Hadjin is a modern town, inhabited solely by Armenians. It is in a great hole in the mountains, many hundred feet below the level of the surrounding country. Every available spot is occupied by a house, and we could not even find a place large enough for our camp. Hadjin is a seat of the American missionaries, whose hospitality we enjoyed during our stay, and whom we left with many regrets.

July 22. Hadjin to a point west of Kilissedjik, 7 h. 6 m. We ascend from Hadjin to the plateau in 1 h. 10 m., and in 3 h. 17 m. from Hadjin we reach the bluff of the great cañon of the Seihûn (Sarus). The cañon is fully one thousand feet deep. The bluffs are almost perpendicular, so that, as one stands on the edge and looks down, it seems scarcely possible for a living being to descend and ascend; yet it may be done. The descent from the top of the bluff to the river bed occupied 37 m. The ascent of the eastern bluff occupied 41 m. Leaving the eastern bluff we travel for 2 h. 35 m. in the direction of Kilissedjik, which point we had hoped to reach, but finding ourselves hopelessly lost, and night having set in, we encamped.

July 23. From our camp 1 h. 59 m. west of Kilissedjik to Göksün, 5 h. 8 m. Kilissedjik is simply an Avshar Yaïla. We found here two tombs of the Graeco-Roman period. The country east of the Seihûn is wooded until within a short distance of the plain of Göksün.

I.

ROMAN MILLIARIA AT COCUSSUS.

The Roman milliaria given below are about eight feet high and three feet in diameter at the base, tapering off to a very thick, blunt point at the top. They are accordingly cone-like in shape. The stones are all very rough and unpolished, and the surface is full of elevations and indentations. It is obvious that inscriptions on such a rugged, uneven surface are very difficult to read, and that, without some practical experience in field epigraphy, one would stand before them absolutely helpless. Impressions of such inscriptions are altogether worthless, as trial has proved to me conclusively.

No. 269.

Göksün (Cocussus). Milestone in the western cemetery. Cf. Bulletin de Correspondance Hellénique, 1883, *p.* 145; Ephemeris Epigraphica, 1884, *p.* 36, *No.* 74; *my* Preliminary Report, *p.* 20, *No.* 13. *Copy and impression.*

```
   I M P
   C A E S
   D I V I S E V E R I N E P
   D I V I M A N T O N I N I
5     F I L ·
   M A V R A N T O N I N O
   P I O F E L I C I A V G
   M I L I A R E S T I T V T A
   P E R M V L P O F E L L I
10 V M T H E O D O R V M
   L E G   A V G
   P R     P R
```

$$P \wedge \Gamma$$

Imp(eratori)
Caes(ari),
divi Severi nep(oti),
divi M. Antonini
5　　　fil(io),
M. Aur. Antonino
Pio Felici Aug(usto)
milia restituta
per M. Ulp(ium) Ofelli-
10　um Theodorum
　　　leg(atum) Aug(usti)
pr(o) pr(aetore).

ρλγ΄

Line 1 fin.　The *Bulletin* omits P.
Line 2 fin.　"　"　adds A.
Line 3 fin.　"　"　reads IIIE and omits P.
Line 5.　　"　"　omits entirely.
Line 9 fin.　"　"　reads OKELI.

Compare Nos. 274, 313, 326, 345.
The emperor is Elagabalus.
This is the one hundred and thirty-third milestone. Note that the numerals of all the stones, with the single exception of the one-hundredth, are in Greek.

Nos. 270–271.

Göksün.　Milestone in the western cemetery.　Stone very rough and inscription hard to read. Preliminary Report, *p.* 21, *No.* 14.　*Copy and impression.*

A.

SALUAL
XIMIANO
⌐VI CAES

B.

```
      A N T O N I V S G O R D I A
 5   N V S   O R I L I S I M V S
      E S A R R E S T I T V I T
        P E R C V S P I D I
        A M I N I V M S E
      V E R V M L E G E T P R
10       P R E T O R E M
```

P M A

This is the one hundred and forty-first milestone.

For the restoration of *A*, see Nos. 318, 323, etc.

For the restoration of *B*, see Nos. 302, 315, and *C.I.L.* VIII. 10342, 10343, 10365.

A mate to this inscription was copied by Mr. Ramsay about six miles to the northeast of Comana, and was published by Mr. Waddington in the *Bulletin de Correspondance Hellénique*, 1883, p. 144, from which it was transferred to *Ephemeris Epigraphica*, 1884, p. 37, No. 77. Mr. Waddington suggests that the fragmentary condition of these inscriptions is probably due to the shortness of the reign of the emperors Pupienus and Balbinus, which, lasting only three months, was no doubt already a thing of the past before the repairs of the roads were completed by the legate Severus. It seems clear that Severus had already caused the inscriptions of Pupienus and Balbinus Augusti and Gordianus Caesar to be engraved on the stones when the news of the deaths of the emperors reached him. Then before putting the milestones in place he caused the names of Pupienus and Balbinus to be erased [not because the names of the emperors had been abolished, but solely for the sake of historical accuracy], changing RESTITVERVNT to RESTITVIT, but leaving to Gordianus III. the title of *Nobilissimus Caesar*, notwithstanding the fact that he was now emperor. Professor Mommsen, in *Ephemeris Epigraphica*, 1884, p. 37, No. 77, calls attention to the erasure of the names of these two emperors in Britain in *C.I.L.* VII. 510 : *Deleta autem sunt omnino non iussu Gordiani, sed errore provincialium longe a turbis illis remotum.*

Mr. Waddington points out further that after the death of Maximinus a certain *Cuspidius Celerinus* proposed to the senate to confer
the imperial purple upon Pupienus and Balbinus (Capitolinus, *vita
Maximini*, 26). It is not improbable, therefore, that our legate,
Cuspidius Flaminius Severus, was a son or other family connection
of *Cuspidius Celerinus*, and that the province of Cappadocia was
bestowed upon him in return for the services rendered by his father.

At a later date the inscription of Diocletian-Maximian-Constantius-Galerius was incised in the place made vacant by the
erasure of the names and titles of Pupienus and Balbinus. Only a
fragment of this inscription now remains, but it is sufficient to enable
one to restore it with certainty.

A.

[Imp(eratoribus) Caes(aribus)
Diocletiano et M.
Aur(elio) Val(erio) Maximiano
P(iis) F(elicibus) Invi(ctis) Aug(ustis) et
Flavi(o) Val(erio) Constantio
et G]al(erio) Val(erio)
[Ma]ximiano
[nob](ilissimis) Caes(aribus).

B.

[Imp(erator) Caes(ar)
M. Clodius Pupienus
Maximus et Imp(erator)
Caes(ar) D. Caelius
Calvinus Balbinus
Pii Fel(ices) Aug(usti) et M.]
Antonius Gordia-
nu[s n]obilissimus
[Ca]e[s]ar restitu(erun)t
per Cuspidi-
[um Fla]minium Se-
verum leg(atum) et [pro-]
pretorem.

ρμα´

Nos. 272–273.

Göksün. Milestone in the southern cemetery. See my
Preliminary Report, *p.* 21 *No.* 15. *Copy.*

I M P

CAESARCIVL
VERVSMAXIMINV
▨▨▨CAESSNC▨▨▨
5   GAIO\IA▨▨▨L I▨▨
D L▨▨▨LE IIANO
ETINVICTO▨▨▨V▨▨
NOBILISSIMVSCAESAR
VIASETPONTESVETV
10   TATECONLABSASRES
TITVERV I▨▨T
P E R▨▨▨▨▨▨▨▨▨▨▨
▨▨▨▨▨▨▨▨▨▨LEG
AVGG   PR   PR
15   XII   P M A

This is the one hundred and forty-first milestone.

This stone has two inscriptions, the one engraved on top of the
other. The *restitutores* of roads and bridges during the reign of
Diocletian-Maximian under *C. Julius Flaccus Aelianus* made use of
the old *Milliaria*, and caused inscriptions of Diocletian-Maximian to
be engraved on them without any regard for the already existing
inscriptions. Thus, as in this inscription and in others below, two
or even three inscriptions are found so mixed up that it requires
both patience and ingenuity to disentangle them.

To inscription *A* belong lines 1–3 and 8–15 inclusive, as well as
the ET at the beginning of line 7. After this ET there followed in
the original inscription of the Maximini the name of C. Julius Verus

Maximinus, the younger, which was afterwards erased. The *restitutores* of Diocletian-Maximian utilized the space thus made vacant
for their own purposes, inserting INVICTO, etc.

Inscription *A* originally read as follows :

> Imp(erator)
> Caesar C. Jul(ius)
> Verus Maximinu[s]
> [Pius Felix Aug(ustus)
> trib(uniciae) pot(estatis) (V?)]
> et [C. Jul(ius) Verus Maximinus]
> nobilissimus Caesar
> vias et pontes vetu-
> [s]tate conlabsas res-
> titueru[n]t per [Licinium
> Serenianum] leg(atum)
> Aug(ustorum) pr(o) pr(aetore).
>
> ρμα′

Compare Nos. 293, 309.

For a restoration of the fragmentary inscription *B*, which is contained in the lines 4–7 inclusive, compare Nos. 270, 288, 301, 318,
323, 324, 327.

Nos. 274–275.

*Göksün. Milliarium in the southern cemetery. See my
Preliminary Report, p. 22, No. 16. Copy.*

```
IMPCAESAR
DIVISEVERI
NEPDIVIM
A  ITONINIFIL
5 MAVRANTON
NOI
            FE
LICIAVG
        SIS
```

10 M I L I A R E
    S T I T V T A P E R
    M V L P O F E L
    L I V M T H
14 ▨▨O P▨▨▨
    ▨▨▨▨▨▨▨

This inscription must be divided into two, of which *A* is contained
in lines 1–6 and 10–14 inclusive, and is preserved almost entire.
Inscription *B*, lines 7–9, is so fragmentary that a restoration cannot
be attempted.

A.

Imp(eratori) Caesar(i),
    divi Severi
nep(oti), divi M(arci)
[An]tonini fil(io),
M. Aur(elio) [A]nton[i]-
no [Pio Felic(i) Aug(usto)]
milia re-
stituta per
M. Ulp(ium) Ofel-
lium Th[e]-
[od]o[rum leg(atum)
Aug(usti) pr(o) pr(aetore)].

No. 276.

*Göksün. Milliarium in the southern cemetery. Badly worn
and wholly illegible with the exception of a few letters and
the numerals. See* Preliminary Report, *p. 22, No. 17. Copy.*

·   L E G
      PR PR
      P Λ

It is the one hundred and thirtieth milestone from Melitene.

No. 277.

Göksün. Milliarium in the southern cemetery. See
Preliminary Report, *p.* 22, *No.* 18. *Copy.*

```
I M P
▨▨▨▨
▨▨▨▨
A V R
R I B
C O ſ A P P
T E ſ· V E T T V ſ T A
N L A P ſ A ſ R E ſ T I T V
I T Δ P K E
```

The AVR in line 4 is not sufficient to authorize a restoration. It
seems probable, however, that the inscription is that of an emperor
other than those mentioned on known milliaria of Cataonia or Melitene.

The last letters PKE look like numerals [125], but the matter is
not certain, because the letters are too small and in the wrong place,
if one may judge by the analogy of all the other numbered milliaria
of Cataonia.

No. 278.

Göksün. Quadrangular cippus. Bulletin de Correspondance
Hellénique, 1883, *p.* 146. *Copy and impression.*[1]

```
A Y P H Λ I O I P ꟺ M A
N O C K A I K E Λ C I A N H
A C K Λ H Π I Δ H T ꟺ
A C Y N K P I T ꟺ Y I ꟺ
5  K A I E M A Y T O I C
M N H M H C X A P I N
```

---

[1] Ligatures occur: line 1, PH; line 2, NH.

Αὐρήλιοι ʽΡωμα-
νὸς καὶ Κελσιανὴ
ʼΑσκληπι(ά)δη τῷ
ἀσυνκρίτῳ υἱῷ
5 καὶ ἐμαυτοῖς
μνήμης χάριν.

Line 1. The *Bulletin* reads ΗΛΙΟΙΡΩΜΛΛ.
Line 2.   "   "   reads NOC · · · CΛCIAN.
Line 3.   "   "   reads · C in init.
Line 4.   "   "   reads · NKI'I.
Line 5.   "   "   reads · A in init.
Line 6.   "   "   reads · M in init.

No. 279.

Göksün. Epistyle block in the cemetery. Bulletin de Correspondance Hellénique, 1883, *p.* 147, *No.* 36. *Copy.*[1]

▓MATOΔ ECTATIAMNHMHIONHΛIOΛWPW-⟶
TEYZAПOCEIΓΛYKEPWΔYCMOPOCWKYM▓

[Σῆ]μα τόδε Στατία μνημῆϊον ʽΗλιο(δ)ώρῳ
τεῦξα πόσει γλυκερῷ δύσμορος ὠκυμ[όρῳ].

Line 1 end. The *Bulletin* reads ΛΙΟΛΟ · · ·
Line 2 end.   "   "   reads CKYM · ·

No. 280.

*Göksün. Round column in the eastern cemetery; broken in
two in the centre. Copy.*[2]

---

[1] In line 1 the fourth letter from the end is certainly Λ by error for Δ.
[2] A cross seems to have been erased from the beginning of the first line.

```
▓OPOITHCA▓▓▓▓
  KAIΘEOT▓▓▓▓
   MAPIA▓▓▓▓
```

$$\cancel{N}$$ ▓▓▓▓

ὅροι? τῆς ἁ[γίας?]
καὶ θεοτ[όκου]
Μαρία[ς].

No. 281.

Göksiin. *Quadrangular cippus in the western cemetery.*
Copy.

```
N o Φ Λ H Λ I
O Δ Ш P Ш
T Ш K Y P I Ш
Π A T P Ш N I
Φ Λ · H Λ I Ш N
K A I Φ Λ A C K Λ Ⱶ▓
Π I▓▓Ш T O C
```

Νο? Φ[λ]. Ἡλι-
οδώρῳ
τῷ κυρίῳ
πάτρωνι
Φλ. Ἡλίων
καὶ [Φλ]. Ἀσκλ[η]-
πι[όδ]ωτος.

No. 282.

Göksün. In the Mussafir Oda of an Armenian house. Stele surmounted by a gable. Bulletin de Correspondance Hellénique, 1883, *p.* 147, *No.* 37. *Copy.*[1]

```
† Є Ν Θ Α Κ Α Τ Α Κ Ι Μ Є Ν Є
Γ Ш Θ Є Ο Δ Ο Ρ Ο С Є Ν Є Ι
Θ Є Ο Υ Λ              Ν Α Γ Ν Ο С
                              Τ Є С
        +      [Female bust]      +

5   Α Γ Ι Α Λ Λ Ο С Μ Α
    Ν Α Ν Υ С Η Α Γ Α Π Η
    Τ Η Θ Υ Γ Α Τ Ρ Ι Κ Α Ι Є
    Α Υ Τ Ш Ι
```

Ἔνθα κατακῖμεν ἐ-
γὼ Θεόδορος
θεοῦ [ἀ]ναγνοστες?

5   Ἀγίαλλος Μα-
νᾶ Νύσῃ ἀγαπη-
τῇ θυγατρὶ καὶ ἑ-
αυτωῖ.

Line 1.   The *Bulletin* reads ΕΝ · · · S · · Ρ · · Є.
Line 2.   "   "   reads ΔΠ for ΔΟ.
Line 3.   "   "   reads · ЄΟΥ   ΝΔΓΝ.
Line 4.   "   "   reads ΓCC.
Line 5.   "   "   reads ΑΠΑΛΟCΜΛ.
Line 7.   "   "   reads Ο for Θ, and ΙΑΙϹ for ΙΚΑΙϹ.
Line 8.   "   "   reads ШΥ.

---

[1] The inscription below the bust was carved by a different hand from the one above the bust.

No. 283.

Göksün. Small quadrangular cippus in the eastern cemetery.
Bulletin de Correspondance Hellénique, 1883, *p.* 146,
No. 34. *Copy.*

AYΡΑΛΕ
ΖΑΝΔΡΟC
ΚΕΚΥΡΙΛΑ
ΝΙΚΕΙΑΤШ
5 ΓΛΥΚΥΤΑ
ΤШΗΜШΝ
ΥΙШΜΝΗΜ
ΗCΧΑΡΙΝ

Αὐρ. ᾿Αλέ-
ξανδρος
κὲ Κύριλ(λ)α
Νικείᾳ τῷ
5 γλυκυτά-
τῳ ἡμῶν
υἱῷ μνήμ-
ης χάριν.

Line 1. The *Bulletin* reads ΜΕ for ΛΕ.
Line 2. " " reads Z for Ζ.
Line 6. " " omits Ν at the end, and does not indicate
a break.
Line 7. The *Bulletin* reads Υ · ШΜΜΑ.
Line 8 is omitted entirely by the *Bulletin*.

No. 284.

Göksün. Flat slab with an immense cross now hewn off. In the southern cemetery. Copy.

ΕΝΘΑΚΑΤΑΚΙΤΕWΤΗС
ΜΑΚΑΡΙΑΜΝΗΜΗСΘWΜΑС
ΟΦΙΛΟΧΡΙСΤΟС

Ἔνθα κατακῖτε ὠ τῆς μακαρία(ς) μντήμης Θωμᾶς ὁ Φιλόχριστος.

No. 285.

Göksün. On a large epistyle block in one single line. In the southern cemetery. Copy.

ΕΙΜΕΝΓΑΡΜΑΚΑΡΕСΜΕΡΟΠWΝΚΡΕΙΝΟΝΤΕСΑΝΑССΟΝΦΠΑСΑΝΑΚΟΙΤΙΝΕΗΝΧ

Εἰ μὲν γὰρ μάκαρες μερόπων κρείνοντες ἄνασσον πᾶσαν ἄκοιτιν ἔην Χ

No. 286.

Göksün. Quadrangular block in the southern cemetery. Copy.[1]

MICANTEMϯ
ΕΥΞΕСΥΝΑ⚹W
ΗСΕΝΑΠΕΙΛΗΝ

[1] In line 3, HN arc in ligature.

July 24. Göksün to Tasholuk, o h. 52 m. Tasholuk is the site of an old town. The plain of Göksün (Cocussus) is remarkable both for its exuberant fertility and for its springs and rivers. Immense springs, sufficient in themselves to form a respectable stream, rise on every hand.

July 25. Tasholuk, *via* Deïrmen Deresi, Kiredj Oghlu, and Göksün, to Yalak, 7 h. 32 m.

No. 287.

Deïrmen Deresi. Large quadrangular cippus. Height, 0.60 *m.; width,* 0.54 *m. See my* Preliminary Report, *p.* 19. *Copy.*

```
ΕΠΙΝΕΡΟΥΑΤΡΑΙΑ
ΝΟΥΚΑΙΣΑΡΟΣΣΕ
ΒΑΣΤΟΥΓΕΡΜΑΝΙ
ΚΟΥΔΑΚΙΚΟΥΕΤΘ
ΔΙΙΕΠΙΚΑΡΠΙѠ
ΚΑΠΙΤѠΝΤΙΛ
ΛΕΥΣΕΚΤѠΝΙΔΙѠ
ΝΑΝΕΘΗΚΕΝ
```

'Επὶ Νέρουα Τραϊα-
νοῦ Καίσαρος Σε-
βαστοῦ Γερμανι-
κοῦ Δακικοῦ ἔτ(ους) θ′
Διὶ 'Επικαρπίῳ
Καπίτων Τιλ-
λεὺς ἐκ τῶν ἰδίω-
ν ἀνέθηκεν.

The name Τιλλεὺς occurs in an inscription of Comana published in the *Bulletin de Correspondance Hellénique,* 1883, p. 137, where it is compared with Τιλλιβόρας, the brigand (Lucian, *Alexandr.* 2).

This inscription was erected in the ninth year of Trajan.   It informs us that Zeus Epikarpios was worshipped here ; indeed, in so fertile a plain, we should naturally expect to meet with the cult of some god of the harvest.

In 2 h. 47 m. from Tasholuk we regain Göksün, and pass on, going up the valley of the Tölbüzek Su, and reaching Mehemet Beikieui in 1 h. 22 m. from Göksün.

II.

MILLIARIA ON THE ROMAN ROAD FROM COMANA TO COCUSSUS.

On this excursion from Göksün to Yalak, which is only six miles from Shahr, we found a number of milliaria, and thus were enabled to trace the Roman road from Comana to Cocussus in its entire length.

Nos. 288-289.

Mehemet Beikieui, one hour to the northeastward of Göksün. *Milliarium defaced by the action of water.   In the ceme-* *tery.   See my* Preliminary Report, *p. 23, No.* 19.   *Copy.*

```
          I M P
    Y  D I O C L E T I A N O
    |      P C I u \ I I
           A        C
                T I T
              M P
                I E
```

Two inscriptions, the one of Diocletian-Maximian, the other of Elagabalus [perhaps], are hopelessly mixed up together.   The restorations must be the same as in the other inscriptions of those emperors in this series.   Compare No. 323 especially.

Nos. 290–291.

Mehemet Beikieui. In the cemetery. See my. Preliminary
Report, *p.* 23, *No.* 20. *Copy.*

A.   IMPCAES
     ARMARCV
     VLPHILIPPVS
     FELIXINVICTVS
5   VGETMARCVS
     HILIPPVSNOBILISSI
     VSCAESARVIA
     PONTESVETV
     ECONLAPSASR
10 STITVENIPERA
     ONMMEMMIVMHI

B.   ETFUAL
     CONSTANT
     NOBCA
15 S C

The uncial text contains the remnants of two inscriptions. Fragment *A*, comprising lines 1–11 inclusive, is almost complete, the cognomen and titles of the legate alone being wanting. Inscription *B* [lines 12–15 inclusive], while much more fragmentary than *A*, still contains sufficient data to make its restoration certain.

A.

Imp(erator) Caes-
ar Marcu[s
J]ulius Philippus
[Pius] Felix Invictus
[A]ugustus et Marcu[s]
[Jul(ius) P]hilippus nobilissi-
[m]us Caesar via[s]
[et] pontes vetu[s-]

[tat]e conlapsas r[e-]
stituerunt per [An-]
[t]on(iu)m Memmium H[ie-]
[ronem leg(atum) Aug(ustorum)
   pr(o) pr(aetore)].

B.

[Imp(eratoribus) Caes(aribus)
· Diocletiano
et M. Aur. Val. Maximiano
P(iis) F(elicibus) invi(ctis) Aug(ustis)]
et F[(lavio)] Val(erio) Constant[io
et Gal. Val. Maximiano]
nob(ilissimis) Ca[es(aribus)].

The three villages, Mehemet Bei, Mahmud Bei, and Taher Bei, are all inhabited by Circassians. There are two uninscribed milliaria at Mehemet Beikieui. Half an hour south of Kürdkieui there is a milliarium almost entirely buried, and it was impossible for us to unearth it.

No. 292.

Dürdkieui (called Kekli Oghlu *on the old map), four hours to the northward of Göksün. The stone never had numerals. See* Preliminary Report, *p. 24, No. 21. Copy.*

```
          C A E
        A R M A R C V
        P H I L I P P V S P I V S F
        N V I C T V S A V G
  5     R C V S I V L P H I L I P P
        B I L I S S I M V S C A E S
        A S E T P O N T E S V E T
        T E C O N L A P S A S R E S
        R A P E R A N T O N V
 10     M I V M H I E R O N E M
        E G A V G G P R
        P R
```

```
     [Imp(erator)] Caes-
     ar Marcu[s Jul(ius)]
     Philippus Pius F[elix]
     [I]nvictus Aug(ustus) [et]
  5  [Ma]rcus Jul(ius) Philipp[us]
     [no]bilissimus Caes[ar]
     [vi]as et pontes vet[us-]
     [ta]te conlapsas res[titue-]
     [runt] per Anton[i]u[m Mem-]
 10  mium Hieronem
     [l]eg(atum) Aug(ustorum) pr(o)
     pr(aetore).
```

The RA at the beginning of line 9 is problematic.

The name of this legate, Antonius Memmius Hiero, is now known with accuracy from this inscription. It occurred, indeed, in an inscription published in the *Bulletin de Correspondance Hellénique*, 1883, p. 142, No. 30, whence it was inserted in the *Ephemeris Epigraphica*, 1884, p. 38, No. 79, but it was in so fragmentary a condition that it had to be restored by conjecture.

From the · · · · ONEM of that inscription Mr. Waddington conjectured [*Seneci*]*onem*, and suggests that the same name must be restored in an inscription of Tavium, published in the *Bulletin de Correspondance Hellénique*, 1883, p. 26. This inscription was also copied by me. But certainly *Hieronem* must now be read instead of *Senecionem*, and in case the two inscriptions make mention of one and the same person, as seems likely, then his full name is M. Antonius Memmius Hiero.

No. 293.

Kürdkieui. *See* Preliminary Report, *p. 24, No. 22. Copy.*

```
     I M P C A
     E S A R I G A
     I O I V I I o V E
     R O M A
     [uncut space]
```

MINO%PIO
FELICI%AVG
TRIB%P%ΓE
LICINNIVM
SESE IMIAN
Vi·LEG%AVG
PRPR

P H

Imp(eratori) Ca-
esari Ga-
io Ju[li]o Ve-
ro Ma[xi-]
mino Pio
Felici Aug(usto)
trib(unicia) p(otestate) [p]e[r]
Licinnium
Se[ren]ian-
u[m] leg(atum) Aug(usti)
pr(o) pr(aetore).

ρλη′

This is the one hundred and thirty-eighth milestone. There are
also two uninscribed stones at Kürdkieui.

No. 294.

Kürdkieui.   Stele.   Copy.

MAPKEΛΛOC
MENANΔPIΔI
T·HXPHCTH
ΓYNEKIKAI
▓CYNKPITⱲ
▓▓AYTⱲ

Μάρκελλος
Μενανδρίδι
τῇ χρηστῇ
γυνεκὶ καὶ
[ἀ]συνκρίτῳ
[καὶ ἑ]αυτῷ.

A short distance northeast of Kürdkieui the watershed is reached.

No. 295.

*Yalak, two hours from Shahr. In the cemetery. Near it
is a defaced milliarium. See my* Preliminary Report,
p. 25, No. 23. Copy.

A R C
L I P P V S
S S I M V S
S A R V I A S E T P
O N T E S V E T V
S T A T E C O N L
P S A S R░░S T░░I
E R V N
N I V X
I V M
M V C
M

[Imp(erator) Caesar
Marcus Jul(ius) Philippus
Pius Felix Invi(ctus) Aug(ustus) et M-]
arcu[s Jul(ius) Phi-]
lippus [nobili-]
ssimus [Cae-]
sar vias et p-
ontes vetu-

state conl[a-]
psas r[e]st[itu-]
erun[t per Anto-]
niu(m) [Memm-]
ium [Hieronem]
[leg(atum) Aug(ustorum)]
[pr(o) pr(aetore)].

No. 296.

Yalak. In the cemetery. See Preliminary Report, *p.* 25,
No. 24. *Copy.*

C Λ Є S A
L I
O C
  U I D Λ E
    O U
  L C I S Λ
P O N
O N L A P S A S

Possibly this is to be restored as an inscription of Constantinus and Licinius, but the indications are too slight to justify it.

No. 297.

Yalak. In the cemetery. Preliminary Report, *p.* 25,
No. 25. *Copy.*

C I A C Y Π A T O
O C T A C O Δ O Y C
T O I O I   O Y
    N T I C T

P M Δ

. . . .

[δημαρχικῆς ἐξου]σίας ὑπατο[ς τὸ . . .]
[πατὴρ πατρίδ]ος τὰς ὁδοὺς [καὶ]
[γεφύρας . . . .]
[διὰ πρεσβευτοῦ καὶ ἀ]ντιστ[ρατήγου . . .]

. . . .

ρμδ΄

This is the only milliarium with a Greek inscription found by me. I made an impression of the stone, but it has been lost with the exception of the numerals. This is the one hundred and forty-fourth milestone.

No. 298.

Yalak. Quadrangular cippus in the cemetery. Copy and impression.[1]

```
    Χ Α Ι Ρ Ε Τ Ω Ο Ν
    Ο C Ο Τ Υ Μ
    Ρ Ε Ι C Ε Π Ι
    Ρ   Ο Ι Ο   C
5   Π Α Ι Δ Ε C Η Γ Ε Ι
    Ρ Α Ν Μ Ε Μ Ν Η
    Μ Ε Ν Ο Ι W C Α
    Γ Α Θ Ο C
```

χαῖρε . . . . . .

. . . . . .

. . . . . ἐπὶ

. . . . . .

παῖδες ἤγει-
ραν μεμνη-
μένοι ὡς ἀ-
γαθός.

---

[1] The vacant places in line 4 were never incised. In line 6, NH are in ligature.

The *Antonine Itinerary* for the whole Antitauran region seems to be hopelessly confused, and its inconsistencies will perhaps never be satisfactorily explained. On p. 210 we read : —

A Coduzalaba
Comana   XXVI
Siricis   XXIIII

while on page 211 we have the following :—

Item a Caesarea Anazarbo CCXI., sic :
Arasaxa   XXIIII
Coduzalaba   XXIIII
Comana   XXIIII
Siricis   XVI
Cocuso   XXV

Now the milliaria given above show that the Roman road between Comana and Cocussus went, as one would naturally expect, by Yalak, Kürdkieui, and Mehemet Beikieui ; and as the whole distance between Shahr and Göksün is reckoned as eight hours, there is plainly something wrong in the statements of the *Antonine Itinerary*. Both Yalak and Kürdkieui are sites of small ancient towns ; but the most important of these was at Yalak, and at Yalak I am inclined to place Siricae. In that case the *Antonine Itinerary* would be nearer the truth if it were emended to read : —

Comana   XXIIII
Siricis   VI
Cccuso   XV

Let it be noted that this, besides being a direct route, is the only natural road-bed between Comana and Cocussus ; on the northeast lies the Bin Bogha Dagh, and on the southwest the Yuvadja Dagh. It is wholly unreasonable to suppose that the Romans would neglect the only natural road-bed to carry a road over the huge mountains just mentioned.

July 26. Yalak to Mehemet Beikieui, 3 h. 35 m. We return in the direction of Göksün.

July 27. Mehemet Beikieui, *via* Kotchos, to Göksün, 4 h. 37 m.
We traced the Tölbüzek Su to its source, which is about three-quarters
of an hour west of Mehemet Beikieui, at the foot of Yuvadja Dagh.
Here innumerable springs gush from the mountain side, and the
water from them is sufficient to form a large swift river of the purest,
coldest water.

No. 299.

*Kotchos.   On the slope of Yuvadja Dagh, in a cemetery near
a Yaïla, about two hours from Mehemet Beikieui.   A pine-
tree has grown around the stone, the beginning of the inscription being buried in the tree.   Copy.*

```
▓▓▓▓▓▓▓▓▓▓▓▓▓▓▓▓
ΓΛ ▓▓▓▓▓▓▓▓▓▓▓
ΙΔΡΙCΥΜΒΙѠ
▓ΑΝΤΙΧΡΗCΤ
ѠCΚΑΙΑΜΕΜ
ΠΤѠCΜΝΗ
ΜΗCΧΑΡΙΝ
```

. . . .
. . . . . . [τῷ ἀ-]
(ν)δρὶ? συμβιώ-
[σ]αντι χρηστ-
ῶς καὶ ἀμέμ-
πτως μνή-
μης χάριν.

III.

MILLIARIA ON THE ROMAN ROAD FROM COCUSSUS TO
ARABISSUS.

July 28. Göksün to Kanlü Kavak, 2 h. 24 m. The road lies in
the plain.

No. 300.

*Milliarium in an old cemetery by the roadside, forty minutes
to the eastward of Göksün. Another milliarium lies deeply
buried by the side of this one. See* Preliminary Report,
p. 27, No. 26. Copy.

```
R V S
A R A O I A R
P O T I V I I
T I M P C A E S I
////////////////////////////
R E S T I T V E R V N T
A N V M  L E G  P R P R
```

[Imp(erator) Caes(ar) L. Septimius Seve-]
rus [Pius Pertinax Augustus]
Ara[b]i(cus) A[diab(enicus), Parth(icus) Max(imus), Pont(ifex)
                                   Max(imus) trib(uniciae)]
pot[e](statis) VII, [Imp(erator) XI, Co(n)s(ul) III, p(ater)
                                   p(atriae), Proco(n)s(ul) e-]
t Imp(erator) Caes(ar) [M. Aurel. Antoninus Augustus
et P. Septimius Geta, nob(ilissimus) Caesar]
restituerunt [per C. Jul(ium) Flaccum Aeli-]
anum leg(atum) pr(o) pr(aetore).

No. 301.

Ibidem. *See* Preliminary Report, *p.* 27, *No.* 27. *Copy.*

M A X I M I A N
N O b C A E
S S

[Imp(eratoribus) Caes(aribus)
Diocletiano et M. Aur. Val.
Maximiano P(iis) F(elicibus)
Invi(ctis) Aug(ustis) et Fl. Val.
Constantio et Gal. Val.]
Maximian[o]
nob(ilissimis) Caes(aribus).

Nos. 302–304.

Ibidem. *See* Preliminary Report, *p.* 27, *No.* 28. *Copy.*

I M P
M A V
M P E R
C·A·C MAXIM
5 C O A N T O R
G O R L N O 6 C A E
L I C I A V C V T O
R E S T I T S S V N T
P E R C V S P I Δ I M
10 M I N I u M S E V E R V M
C A T V M P O P R A C
Γ O A

At first sight the difficulties of this inscription seem to be insur-
mountable, but they disappear by the help of the elucidations given

above under No. 271. The original inscription was that of Pupienus and Balbinus Augusti and Gordianus Caesar, the close of which is found in lines 8–12 inclusive, and which read originally as follows :

A.

[Imp. Caes. M. Clodius Pupienus
Maximus et Imp. Caesar
D. Caelius Calvinus
Balbinus Pii Felices Augusti
et M. Antonius Gordianus
nobilissimus Caesar]
restit[uerunt]
per C[u]spidium [Fla-]
minium Severum [le-]
[g]atum p(r)o pr[ae]-
to[re].

Then after the erasure of the names of Pupienus and Balbinus a new inscription of Gordianus III., couched in terms different from those of the original inscription, was incised in the place made vacant by the erasure. As in Nos. 271, 316 the closing lines of the first inscription were allowed to stand, notwithstanding the fact that they were out of place both grammatically and historically. The remnants of this inscription are to be sought in lines 3, 5–7 inclusive, and must be restored somewhat as follows :

B.

[I]mper[atori] Caesari Mar-]
c[o A]nto[nio]
Gor[dian]o [Pio Fe-]
lici Augu[s]to.

Lastly, line 4 is almost certainly to be restored as

GALVALMAXIMIANO

and consequently we have before us an inscription of Diocletian-Maximian-Constantius-Galerius Maximianus. To this inscription belong lines 1–2, 4, and the latter part of line 6. It must be restored as follows :

C.

[Impp. Caess. Diocletiano
et M. Aur. Val. Maximiano P(iis) F(elicibus)
Inui(ctis) Aug(ustis) et Fl. Val. Constantio et]
Gal. Val. Maximiano
nob(ilissimis) Caes(aribus).

In this cemetery there is still a fourth milliarium, deeply imbedded.
In a cemetery 1 h. 5 m. east of Göksün there are two more milliaria; one nearly buried, the other erect but illegible. It was impossible for us to get at half-buried stones that were distant from a
village : to raise one out of a hole is generally the work of half a day
for four men in a country where levers are not to be had.

No. 305.

*In an old cemetery by the roadside, one hour and forty
minutes to the eastward of Göksün. See* Preliminary
Report, *p.* 28, *No.* 29. *Copy.*

PERMEMM

If the name of the legate was Antonius Memmius Hiero, then the
inscription stood in the name of the Philippi.

No. 306.

Ibidem. Erect. See Preliminary Report, *p.* 28, *No.* 30. *Copy.*

T MAXI ONTIM
M XII COS IIII P IBO
IMI AVBE ANTONINYS
I T
PEPHYLIYMFLACICYMIAEWAYM EO

[Imp(erator) Caes(ar) L. Septimius Severus
Pius Pertinax Aug(ustus), Arab(icus), Adiab(enicus),
Par]t[h(icus)] Maxi[(mus), P]onti(fex) M[ax(imus), trib(uniciae)
　　　　　　　　　　　　　　　pot(estatis)- ?]
[I]m(perator) XII, Co(n)s(ul) III[I], p(ater) p(atriae), Proco(n)s(ul),
[et] Im[p](erator) [C(aesar) M. A]u[r]e[l.] [A]ntoninus [Aug(ustus)
et P. Septimius Ge]t[a nob(ilissimus) Caes(ar) restituerunt]
pe[r] (C). (I)ulium Flac(i)cum(i) Ae[li]a[num leg(atum)
　　　　　　pr(o) pr(aetore)].

No. 307.

Ibidem. Erect. See Preliminary Report, *p.* 28, *No.* 31. *Copy.*

I M
L S E
P I V S
P A R T I
I M P X I
U A V R
E T L I S E
P E R C I V L

Im[p(erator) Caes(ar)]
L. Se[ptimius Severus]
Pius [Pertinax Aug(ustus), Arab(icus), Adiab(enicus),
Part[h(icus) Max(imus), Pont(ifex) Max(imus), trib(uniciae)
　　　　　　　　　　　　pot(estatis) VI],
Imp(erator) XI, [Co(n)s(ul) III, p(ater) p(atriae), Proco(n)s(ul) et
Imp. Caes. M]. Aur[el(ius) Antoninus Aug(ustus)
et [P.] Se[ptimius Geta nob(ilissimus) Caes(ar) restituerunt]
per C. Iul(ium) [Flaccum Aelianum leg(atum)
　　　　　　pr(o) pr(aetore)].

No. 308.

Ibidem. Erect. See Preliminary Report, *p.* 28, *No.* 32. *Copy.*

```
d I O C L E T I A
E T     A  N
```
```
I T   A  U
C O N S T A   T I O
E T C A I U M
M A X I M I A N O
N           P R
```

[Imp(eratoribus) Caes(aribus)]
  Diocletia[no]
et [M. Aurel(io) Val(erio) Maximiano
Piis Felici(bus) Invi(ctis) Aug(ustis)
e]t [Fl]a[v]i(o) V[al(erio)]
  Consta[n]tio
et [G]a[l]. V[(al)].
  Maximiano
[nobb. Caess.].

No. 309.

Ibidem. Erect. See Preliminary Report, *p.* 29, *No.* 33. *Copy.*

```
            A E S
          C   O I V L
      R O M A X I M I
      P I O F E L I C I A
V G T R I B P P E R L I
C I N N I V M S E R E N I
A N V N L E G A V G
    P R P R
        P K B
```

[Imp(eratori) C]aes(ari)
C[ai]o Iul[io]
[Ve]ro Maximi[no]
Pio Felici A-
ug(usto) trib(unicia) p(otestate) per Licinnium Sereni-
anum leg(atum) Aug(usti)
　　pr(o) pr(aetore).

ρκβ′

This is the one hundred and twenty-second milestone from Melitene.

In the cemetery by the roadside 20 m. southeast of Kanlü Kavak we found no less than twenty-six milliaria, many of which were never inscribed. The inscribed stones cost us a day and a half of hard work in deciphering and copying the inscriptions.

No. 310.

Kanlü Kavak. Milliarium in the old cemetery which is on the main road leading from Göksün to Yarpuz, and about twenty minutes to the southeastward of Kanlü Kavak. See Preliminary Report, *p.* 29, *No.* 34. *Copy.*

M P
S P I
I C T V
A R C V S
N O B I L I S S I M
⌐ S A R V I A S E T P O
T▨S V E T V S T A T E
C O N⌐A P S A S R E S T I T V E ·
P E R A N T O N I V M M▨C
M I V M H I E R O N E M
L E G A V G
P R P R

[I]mp. [Caes.]
[M. Iuliu]s P[hilippus Pius]
[Felix Inv]ictu[s Aug(ustus)]
[et M]arcus [Iulius Philippus]
     nobilissim[us]
[Cae]sar vias et po-
[n]t[e]s vetustate
conlapsas restitue[runt]
per Antonium M[em-]
mium Hieronem
leg(atum) Aug(ustorum)
     pr(o) pr(aetore).

No. 311.

Kanlü Kavak.   Ibidem.   See Preliminary Report, *p.* 30,
No. 35. *Copy.*

V

M

Є

5   I C I

 ˙V N I

  I O N

P R P R

P K

This is the one hundred and twentieth milestone from Melitene.

The ON in line 7 seems to indicate that the name of the legate was Antonius Memmius Hiero, but it is not advisable to restore the inscription on the strength of these two letters alone.

No. 312.

Kanlü Kavak.   *See* Preliminary Report, *p.* 30, *No.* 36. *Copy.*[1]

```
     ⫫ N ⫻⫻⫻⫻⫻ N O
     ⫻⫻⫻⫻L I S S I M O C A S A
     C A T C L E M E N T
     C R C R C R O V I ᴎ C I A
5         I M P
```

P K E

. . . .

. . . . . . [nobi]lissimo Ca[e]sa[ri]

[per] Cat(ium) Clement[em]

[leg(atum) Aug(ustorum) p]r(o) [p]r(aetore) [p]rovi[n]cia[e]

$\rho\kappa[\epsilon']$

No. 313.

Kanlü Kavak.   *See* Preliminary Report, *p.* 30, *No.* 37.   *Copy.*

```
         I M P
         D I V I S E V E R I
         N E P D I V I M A N
         T O N I N I F I L
         M A V R        ·
         N O P I O F E L I C I
            A V G
         M I L I A R E S T I T V T A
         M⫻⫻⫻P O F E L L I V M
       T H E O D O R V M
         L E G A V G P R P R
```

M ⫻ K I

---

[1] I have a note to the effect that I was doubtful while in the presence of the stone as to whether line 5 should read IMP or IHP.

Imp. [Caes(ari)],
divi Severi
nep(oti), divi M. An-
tonini fil(io),
M. [A]ur(elio) [Antoni-]
no Pio Felici
    Aug(usto)
milia restituta [per]
M. [Ulp]. Ofellium
Theodorum
leg(atum) Aug(usti) pr(o) pr(aetore).
M(ίλια) [ρ κ[ε′ or η′]

This must be the one hundred and twenty-fifth or else the one hundred and twenty-eighth milestone from Melitene, as only E or H can be restored as the missing numeral.

No. 314.

Kanlü Kavak. *See* Preliminary Report, *p.* 31, *No.* 38. *Copy.*

P R P R

Nos. 315–316.

Kanlü Kavak. *Sec* Preliminary Report, *p.* 31, *No.* 39. *Copy and impression.*

I M P
C A E S A R I M A .
R C O Λ N T O N I
O Ç O R Δ I A N O P I
5 O F E L I C I A V G V S
T O R E S T I T V E R V
N T P E R C V S P I Δ
I V M F S A M I N I
V M S E V E R V M
10 L E G A T V M P R O P
P Λ E Ν T O R E M

In the light of Nos. 271, 304, this inscription becomes plain, and falls into two inscriptions. The name of the legate Cuspidius Flaminius Severus fortunately is preserved here in full, and from it we learn that the original inscription was one of Pupienus and Balbinus Augusti and Gordianus Caesar. What is now left of it is contained in the lines 6-11 inclusive, with exception of the TO at the beginning of line 6, which belongs to the second inscription. The original inscription read as follows :

A.

[Imp(erator) Caesar M.
Clodius Pupienus Maximus
et Imp. Caes. D. Caelius
Calvinus Balbinus
Pii Felic(es) Aug(usti) et
M. Antonius Gordianus
nob(ilissimus) Caes(ar)]
restitueru- .
nt per Cuspid-
ium F(l)amini-
um Severum
legatum prop-
[ra]etorem.

After the erasure of the names of Pupienus and Balbinus the new inscription of Gordianus III. Augustus was incised, and is preserved intact in lines 1-5 inclusive, to which must be added the TO at the beginning of line 6. It reads :

B.

Imp(eratori)
Caesari Ma-
rco [A]ntoni-
o Gordiano Pi-
o Felici Augus-
to.

No. 317.

Kanlü Kavak. Two inscriptions are so inscribed on and over each other that it is perhaps impossible to disentangle them, but the lines given below can be read. See Preliminary Report, *p.* 31, *No.* 40. *Copy.*

```
I M ˉ
C A E S M A R C V S
I V L P H I L I P P V S
P I V S F E L I X
```

Im[p](erator)
Caes(ar) Marcus
Iulius Philippus
Pius Felix
5 [Invi(ctus) Aug(ustus) et
Marcus Iulius Philippus
nob(ilissimus) Caesar
vias et pontes vetus-
tate conlapsas restitu-
10 erunt per Antonium Memmium
Hieronem leg(atum) Aug(ustorum)
pr(o) pr(aetore)].

Nos. 318–319.

Kanlü Kavak. See Preliminary Report, *p.* 31, *No.* 41. *Copy.*

A.
```
I M P P▨▨▨▨▨
▨▨▨O C▨▨▨T I A I
E T ᴎA L T I U A L
M A X I M I A N O.
5 P P F F I N U A U G
E T F I Λ ᷽᷽᷽A I
```

C O N S T A N T I O

E T ░░░░ A I

C A E S

10 M A X I M I A N O

S E V E R V S

[A blank, apparently uncut space.]

B.   H M A X P O N T M A X T R I B P O T V I O

I P X I C O S I I P P P R O C O S E T I M P C A E S

.M A V R E L · A N T O N I

N V S A V G [name erased]

5 [name erased] T I T V E R V N T

P E R C · I V L I V M F L A C

C V M A E L I A N V M L E G P R P R

I have a note to the effect that lines 8, 9, 10 of *A* are written together, and are so mixed up as to be exceedingly doubtful.

By some mistake, which I am unable to explain, line 11 of inscription *A* does not appear in the *Preliminary Report.*

It must be noted especially that lines 9 and 11 of inscription *A* certainly belong to inscription *B*, which see below.

A.

Imp(eratoribus) Caes(aribus)

[Di]oc[le]tia[no]

et [M]. A[ur(elio)] Val(erio)

Maximiano

P(iis) F(elicibus) Inv(ictis) Aug(ustis)

et F[la]vi(o) Va[l](erio)

Constantio

et [Gal](erio) [V]a[l](erio)

Maximiano

[nob(ilissimis) Caes(aribus)].

B.

[Imp(erator)] Caes[ar
L. Septimius] Severus
[Pius Pertinax Aug(ustus), Arab(icus), Adiab(enicus),
Parth](icus) Max(imus), Pont(ifex) Max(imus), trib(uniciae)
pot(estatis) VI,
I(m)p(erator) XI, Co(n)s(ul) II[I], p(ater) p(atriae),
Proco(n)s(ul) et Imp(erator) Caes(ar)
M. Aurel(ius) Antoni-
nus Aug(ustus) [et P. Septimius Geta
nob(ilissimus) Caesar res]tituerunt
per C. Iulium Flac-
cum Aelianum leg(atum) pr(o) pr(aetore).

Nos. 320–321.

Kanlü Kavak. *See* Preliminary Report, *p.* 32, *No.* 42.
Copy and impression.

```
        I M P P
      d I O C L E T I A N O
                T
  I M P C A E S M A V R U A L
              MAXIMIANO
  L S E P T I M I V S S E V E R V S A V G
               PPLE INV
 5 P I V S P E R T I N A X A V G A R A B I A D I A B
  P A R T H M A X P O N T M A X T R I B T I I O T V I
                       OICT
  I M P X I C O S I I I P P P R O C O S E T I M P C A E S
                          ET CAIVA
  M A V R E L A N T O N I N V S A V G N O
  E T L S E P T░░░░V S N O b b C A E S S░░E S T I T V E R V N T
10 P E R C · I V L I V M F L A C C V M A E L I A N V M L E G P R P R
```

The two inscriptions are badly confused on the stone, inasmuch
as the later inscription of Diocletian-Maximian-Constantius-Galerius
Maximian (*B*) has been incised over the older inscription of L.

Septimius Severus (*A*) in such fashion as to make the whole unintelligible at the first glance. It is impossible to present the inscriptions accurately in uncial text, but I have tried to give at least an approximate idea of the truth.

Let us endeavor to disentangle the inscriptions !

Lines 1 and 2 belong wholly to *B*. Of line 3 IMPCAES belongs to *A*, and MAVRVAL to *B*. Besides this a T was inserted after the E of CAES, and the ET thus obtained belongs to *B*.

MAXIMIANO was inserted between lines 3 and 4, and belongs to *B*. All of line 4 belongs to *A* excepting the closing AVG, which belongs to *B*, and follows the PP[F]E⧸⧸INV which is inserted between lines 4 and 5. In this line the A of AVG has been so carved as to resemble a ligature with the closing S of SEVERVS.

Lines 5, 6, 7 belong wholly to *A*, only perhaps at the close of line 6 a disturbing effect has been produced by the incision of something belonging to *B*. The letters between lines 6–7 and 7–8 belong to *B*. The NO at the close of line 8 probably originally followed the name of P. Septimius Geta in line 9, that being the only theory upon which I can account for its presence, which is certified by the impression.

Lines 9, 10 belong to *A*, with the exception of the NObbCAESS in the middle of line 9, which belongs to *B*, and was incised in the place made vacant by the erasure of the name of Geta.

After these preliminary explanations it will be clear that the inscriptions must be restored to read as follows :

A.

Imp(erator) Caes(ar)

L. Septimius Severus

Pius Pertinax Aug(ustus), Arab(icus), Adiab(enicus),

Parth(icus) Max(imus), Pont(ifex) Max(imus), trib(uniciae),

⌊pot⌋(estatis) VI,

Imp(erator) XI, Co(n)s(ul) III, p(ater) p(atriae), Proco(n)s(ul),

et Imp(erator) Caes(ar)

M. Aurel(ius) Antoninus Aug(ustus)

et (P). Sept⌈imi⌉us [Geta] no⌈b⌉(ilissimus) [Caes(ar) r]estituerunt

per C. Iulium Flaccum Aelianum leg(atum) pr(o) pr(aetore).

B.

Impp. [Caess.]
Diocletiano
et M. Aur. Val.
Maximiano
5   P(iis) F(elicibus) Inv(ictis) Aug(ustis)
[et Flavi(o) Val(erio)
Constantio]
et [G]a[l]. Va[l].
[Maximiano]
10   nobb. Caess.

Inscription *A* belongs to the year 203 A.D., and inscription *B* falls between 293 A.D. [the year in which Constantius and Galerius were made Caesares] and 305 A.D. [the year in which the Augusti Diocletian and Maximian abdicated].

Nos. 322–323.

Kanlü Kavak. See Preliminary Report, *p.* 32, *Nos.* 43 *and* 44. *Copy and impression of A. Copy of B.*

A.

I M
R C A E S M A
R C V S I V L P
H I L I P P V S P I V
5   S E E L I X I N V I
C T V S A V G E
T M A R C V S
I V L I V S P H I
L I P P V S N O
10 B I L I S S I M V S
C A E S A R V I A
S E T P O N T E

▨▨ E T V S T A T
▨▨▨ N L A P S A S
15   ▨▨▨▨▨ V E R
▨▨▨▨▨▨ P A

On the other side of the stone.

B.

I M P P E L (SV Λ ı
d I O C L E T I A N O
E T M A U R U A L
M A X I M (A N O
5 P P F F I N U I A U Ç
U I U A I
C O N S T A N T I O
E T C A ▨▨A▨E
M A X I M I A N O
10 N O b b C A E S S

A.

Im-
(p)(erator) Caes(ar) Ma-
rcus Iul(ius) P-
hilippus Piu-
5 s (F)elix Invi-
ctus Aug(ustus) e-
t Marcus
Iulius Phi-
lippus no-
10 bilissimus
Caesar via-
s et ponte-
[s v]etustat-
[e co]nlapsas
15 [restit]uer-

[unt per] A-
[ntonium
Memmium
Hieronem
20   leg(atum) Aug(ustorum)
        pr(o) pr(aetore)].

B.

Impp. [Caess.]
Diocletiano
et M. Aur. Val.
Maximiano
5   P(iis) F(elicibus) Invi(ctis) Aug(ustis)
    [et Fla]vi(o) Va[l](erio)
    Constantio
    et [G]a[l. V]a[l].
    Maximiano
10   nobb. Caess.

No. 324.

Kanlü Kavak. See Preliminary Report, *p.* 33, *No.* 45. *Copy.*

```
         A X
 5   P P F F I N V I A V G
     E T F L A V I V A L
     C O N S T A N T I O
     E T C U A L E N
     M A X I M I A N O
10   N O b b C A E S S
     P
```

[Impp. Caess.
Diocletiano
et M. Aur. Val.
M]ax[imiano]
5   P(iis) F(elicibus) Invi(ctis) Aug(ustis)

> et Flavi(o) Val.
> Constantio
> et [G](al). Vale[ri](o)
> Maximiano
> 10   nobb. Caess.
>
> ρ? . . ´

Nos. 325-327.

Ḳanlü Kavak. Milliarium with three inscriptions inscribed on and over each other. After much labor I succeeded in disentangling them. See Preliminary Report, *p.* 34, *Nos.* 46–48. *Copy and impression.*

A.

```
        I M P C A
    E S A R G V I V I V S T R E B O
    N G A L L V S E T I M P C A E S
    A R G V I V I V S   V E L D V M I
  5 N I A N V S   V O L V S I A N V
    P,I I F E L I C I N V I C T I A V G G V I A S
    E T P O N T E S V E T V S T A T E C O N
    L A P S A S R E S T I T V E R V N T P E R A
    V E R G I L I V M M A X I M V M V C
 10     V G G P R P R
```

B.

```
  8 R E S T I T V T A
      E R M V L P
 10 O F E L L I V M
  .  T H E O D O R V
    M L E G A V G
      P R P R
```

M

C.

IMPPCC
. dIOCLETIANO
ETMAVRULI
MAXIMIANO
5 PPFFINVIAUG
ETFᏞAU!VAL
CONSTANTIO
ETCAIUAL
MAXIMIANO
10 NOᏏᏏCΛESS

A.

Imp(erator) Ca-
esar G. Vi[v]ius Trebo-
n(ianus) Gallus et Imp(erator) Caesar G. Vivius Veldumi-
5 nianus Volusianu[s]
Pii Felic(es) Invicti Aug(usti) vias
et pontes vetustate con-
lapsas restituerunt per A(ulum)
Vergilium Maximum v(irum) c(larissimum)
10 [leg](atum) [A]ug(ustorum) pr(o) pr(aetore).

B.

[Imp(eratori) Caes(ari),
divi Severi nep(oti),
divi M. Antonini
fil(io),
5 M. Aur. Antonino
Pio Felici Aug(usto)
milia]
restituta
[p]er M. Ulp(ium)

10   Ofellium
     Theodoru-
     m leg(atum) Aug(usti)
     pr(o) pr(aetore).
     [ρ]μ.?

C.

     Imp(eratoribus) C(aesaribus)
     Diocletiano
     et M. Aur. V(a)l.
     Maximiano
5   P(iis) F(elicibus) Invi(ctis) Aug(ustis)
     et F[l]av[i](o) Val.
     Constantio
     et [G]a[l]. Val.
     Maximiano
10   nob(ilissimis) Caes(aribus).

No. 328.

*Kanlü Kavak. A quadrangular cippus in the cemetery has
a defaced inscription, of which only* ΓΛΥΚΥΤΑΤШ *is to be
deciphered.*

July 30. Kanlü Kavak, *via* Aristülü and Kizildjik, to Kayadibi,
6 h. 37 m. Between Kanlü Kavak and Yarpuz no milliaria were
found. Indeed, all seem to have been transported from this whole
region to the cemetery of Kanlü Kavak to serve as tombstones. The
road of to-day traverses a rough and inhospitable country, but at
Kayadibi the plain is again reached.

July 31. Kayadibi, *via* Nadin and Altash, to Yarpuz, 6 h. 56 m.
The Göksün Su was crossed twice to-day; it is a large stream here.
Leaving Ertchin we cross the low Atlas Dagh to Yarpuz.

No. 329.

Yarpuz (Arabissus). In the cemetery. So superscribed as to be hopelessly illegible. See Preliminary Report, *p.* 35, *No.* 50. *Copy.*

NOBILISSIMI
CAES

No. 330.

Yarpuz. In the cemetery; erect; illegible. See Preliminary Report, *p.* 36, *No.* 51. *Copy.*

CONLAP

No. 331.

Yarpuz. Used as a step in the Djami. It is much worn, and the letters are very uncertain. Copy.

TAVITSETAN
NUSXXECIT
REPOIT
ESTIESCU
IUTCUT
ITAINCN
NKR
IRIIETEI
BITASTEOI
LVMCIII
PERINTT
ASVAB
ERITWOSVIT

No. 332.

Yarpuz.   Stele with immense cross in the Armenian church.   Copy.

```
†ΟΤΑΣΔΩΡΕΑΣΤΟΥΘΥΠΛΟΥΣΙΑΣ
ΔΕΖΑΜΕΝΟΣΚΤΟΝΠΟΛΥΜΟΧΘΟΝ
ΒΙΟΝΡΑΟΤΕΡΟΝΠΑΡΑΔΡΑΜΩΝ:ΕΝ
ΘΑΔΕΚΑΤΑΚΙΜΕΙΦΙΛΑΓΡΙΟΣΕΙΣΤΗΝ
5  ΤΟΥΟΙΚΕΙΟΥΠΡΟΣΤΑΤΟΥΚΑΤΑΦΥΓΩΝ
ΑΝΤΙΛΗΨΙΝ†
```

> Ὁ τὰς δωρεὰς τοῦ θ(εο)ῦ πλουσίας
> δεξάμενος κ(ὲ) τὸν πολύμοχθον
> βίον ῥᾳότερον παραδραμών ἐν-
> θάδε κατακῖμει Φιλάγριος εἰς τὴν
> 5   τοῦ οἰκείου προστάτου καταφυγὼν
> ἀντίληψιν.

Line 4.   κατακῖμει stands for κατακεῖμαι.

No. 333.

Yarpuz.   On a sarcophagus in the court of the Armenian church.   A large cross divides the inscription in two.   On either side of the upright bar of the cross is represented a peafowl.   Copy.

```
ΩΕΒΛΑΒΕΣΣΤΑ
ΟΣΠΡΕΣΒΥΣΤΕΡΟΣ
ΑΖΜΑΝΤΟΣΕΝΘΑ
ΕΚΑΤΑΚΙΤΕ
```

> Ὠ ἐβλαβέσ(σ)τατ-
> ος πρεσβύστερος

*Ἄζμαντος ἐνθά[δ.]
ε κατακῖτε.

Line 1 stands for ὁ εὐλαβέστατος.
Line 2. The form πρεσβύτερος occurs here for the first time, so far as I can find out.
Line 3. *Ἄζμαντος is a native name hitherto unknown.
Line 4. κατακῖτε for κατακεῖται.

No. 334.

Yarpuz. In the wall of the Djami. Copy.

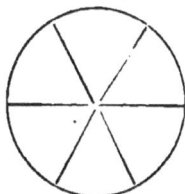

```
      OY              KЄBOHΘH
  ΛΞITON            ΔΟΥΛΟΝC
    ΙΗΞΟ             IΟΥCΤΙΝΟ
   ΔΟΡЄ
   ΟΙΚΟΥ
```

K(ύρι)ε βοήθη [τὸν]
δοῦλόν σ[ου]
᾿Ιουστῖνο[ν].

No. 335.

Yarpuz. Stele in the wall of a house. Copy.

ΜΑΑΤΙΝΑΤѠ
ΑΝΔΡΙΜΝΗ
ΜΗCΧΑΡΙΝ

Μᾶ ᾿Ατινάτῳ
ἀνδρὶ μνή-
μης χάρω.

No. 336.

Yarpuz. Stele with large cross in an Armenian house. Copy.

```
▨▨▨ЄΠΑΥϹΑΤΟΗΛΟΥ
▨▨▨ΟΥϚЄΟΥΜΑΡΙΑ
▨▨▨‾ΗΜ̇ΟΚΤШΒΡΙШΚ̅Δ̅
▨▨▨ΡΑϹΚЄΥΗ†
```

["Ενθα?] ἐπαύσατο ἡ [δ]ού-

[λη τ]οῦ θεοῦ Μαρία

. . . . η μη(νὶ) 'Οκτωβρίῳ κδ'

[ἡμέρᾳ Πα]ρασκευῇ.

Καταπαύω is used intransitively in the Septuagint version of Genesis ii. 2.

Arabissus, now Yarpuz, was once an important place, to judge by the remains still extant, which, however, are mostly Christian.

The afternoon of this day was spent in an excursion to Ziyaret Serai, 1 h. 10 m. east of Yarpuz.

Ziyaret Serai is a Seldjukian palace or villa, now falling into decay.

No. 337.

In the old cemetery between Emirli and Ziyaret Serai.
See Preliminary Report, *p.* 35, *No.* 49. *Copy.*

```
     IMPTRIBPOTVIA
        ET        NTE
     SR    STITVERVN
     CIVLI       OCI .
  5     AVG       PR
            C
```

The name of the legate is probably C. Julius Flaccus Aelianus, and consequently the inscription belongs to Septimius Severus. Still the data are too insignificant to make this certain.

All the other milestones copied by me have Greek numerals. This one alone having the Latin C, it being the one hundredth milestone from Melitene.

No. 338.

Inscribed on a panel smoothed out on the face of the rock on the mountain side, south of and immediately above the cemetery mentioned in connection with the last inscription. There is no means of telling how much of the panel has been broken away.

```
        ΑΟΥΙϹΟϹ
        ΟΥΜΑΡΙΑ.
        ΚΑΙΠΑΠΕΙ
        ΥϹΗϹ
  5     ΟΔΟΥ
        ϙϹΕΤΟΥϹ
        ΕΚΤΙϹ
        ⅃ΕΛΟϹ
        Ϲ Ѡ
  10    ΙΕΙΡΙΟΥ
        ΓΟΡΟϹ
        ΚΗΤΟΥ
```

. . . . ου Μαρία
. . . . καὶ Παπει
. . . .
. . . . ὁδοῦ
. . . . ἔτους
. . . . ἔκτισ-
α . . . . ἄμπ]ελος
. . . . σω

At Yalak the one hundred and forty-fourth milliarium (No. 297) was found. By a glance at the numerals of the milliaria between Yalak and Yarpuz it will be seen that the numerals diminish steadily along this road, a fact which proves conclusively that distances in the Trans-Antitauran region were measured from Melitene as the starting-point.

From Göksün the Göksün Su goes down a narrow valley, and does not flow south of Beirüt Dagh, as it is made to do on the old map constructed from von Moltke's hurried ride.

August 1. We undertook a journey in a northerly direction, with Khurman Kalesi as an objective point. The time from Yarpuz, *via* Khunu and Norshun, to Indjiler was 4 h. 40 m.

No. 339.

Khunu. Quadrangular cippus in the cemetery. Copy.

ΑΓΝΟΤΑΤШΠΟΝ▨
ΡΙΟCΙΗCΧΑΡΙΝΤΟ▨
ΟΙΚΟΥΟΛΥΜΠΟ▨
ΑΝΕCΤΗCΑC

$$\text{'Αγνοτάτῳ Πο[ν ?]}$$
$$\text{ρι ὁσίης χάριν το[ῦ]}$$
$$\text{οἴκου 'Ολύμπο[υ]}$$
$$\text{ἀνέστησα⟨ς ?⟩}$$

Arrived at Indjiler we find that we have lost the road to Khurman Kalesi, and are advised to take a short cut through the mountains. After wandering about in the uninhabited mountains until midnight we reluctantly camped out.

August 2. We left camp at peep of day, and for a wonder found Khurman Kalesi at 4 o'clock A.M. Not being able to find food for man or beast, we had to leave immediately for Tanir. The inscriptions, for which we had undertaken the journey, were found afterwards (Nos. 352–354). The time from Khurman Kalesi, *via* Tanir, Norshun, and Merki, to Yarpuz was six hours. Tanir is the site of an old town; no doubt the name is a corruption of ΠΤΑΝΔΑΡΙΣ.

No. 340.

Merki. Stele. Copy.

A M M H
Z H Θ Ω I
T Ω I Y Ω I
M N H M H C X A
P I N ,

Ἄ[μ]μη
Ζήθωι
τῶι υἱῶι
μνήμης χά-
ριν.

We remained a day in Yarpuz to allow our horses to recruit, and to recruit ourselves.

August 4. Yarpuz to Albistan, 3 h. 56 m.

IV.

MILLIARIA ON THE ROMAN ROAD FROM ARABISSUS TO MELITENE.

No. 341.

In an old cemetery one hour and four minutes east of Yarpuz.
See Preliminary Report, *p.* 36, *No.* 52. *Copy.*

I
' R E S T I T
P E R
C I V L I V M F L A C
C V M A E L I A N V M L E G P R P R

MIL P

[Imp. Caes.

L. Septimius Severus

Pius Pertinax Aug. Arab. Adiab.

Parth. Max. Pont. Max. trib. pot. VI.

Imp. XI, Cos. III. p.p., Procos. et Imp. Caes.

M. Aurel. Antoninus et P.

Septimius Geta nob. Caes.] restit[uerunt]

> per
>
> C. Iulium Flac-
>
> cum Aelianum leg. pr. pr.
>
> Mil(ia) P.

If the P be a Greek numeral, as is likely, then this is another one hundredth milestone, but it is noteworthy that this is the only stone with MIL in Latin.

No. 342.

In an old cemetery one hour and forty minutes east of Yarpuz.
See Preliminary Report, *p.* 36, *No.* 53.   *Copy.*

IMP%CAESAR

AVREL

[Space overwritten.]

%POTEST%COS

UIASETPONT

5   UETTUSTAT

APSASREST

T%

A restoration cannot be attempted on the sole authority of line 2, and the inscription is probably that of an emperor not mentioned on any of the known milliaria of Cataonia.

I had been suffering from fever ever since our disastrous journey to Khurman Kalesi, and here the fever had reached such a pitch that I had to abandon work for to-day. The two milliaria in the cemetery of Isgin (Nos. 343–344) were copied by Mr. Haynes.

No. 343.

Isgin. In the cemetery. Copied by J. H. Haynes. See
Preliminary Report, *p.* 37, *No.* 54.

```
       E R O C O S
  ⌐" T I M P C A E S · M · A V R E L ·
  L
  A N T O N I N V S · A V G
  E T I ◾ S E P T I M I V S
  G E I A C A E S R E S T I T
  V E R V N T · P E P · C · I V L I V M
  E L A C̨ C V M · A E L I A N V M
         L E      P R      P R
```

[Imp. Caes.
L. Septimius Severus
Pius Pertinax Aug. Arab. Adiab.
Parth. Max. Pont. Max. trib. pot. VI
Imp. XI, Cos. III, p.p. P]rocos.
[et] Imp. Caes. M. Aurel.
Antoninus Aug.
et [L]. Septimius
Ge[t]a Caes. restit-
uerunt per C. Iulium
[F]laccum Aelianum
le[g]. pr. pr.

No. 344.

Isgin. In the cemetery. Copied by J. H. Haynes. See
Preliminary Report, *p.* 37, *No.* 55.

```
        · C A E S
    T R I B P O T E S T
    A S R E S T I T
```

Rev. Henry Marden has found a Hittite inscription at Isgin. My excuse for not having found it myself is that I was very ill, and lay in agony in an Oda in Isgin for the greater part of the day.

We found nine milliaria at Albistan, some of which were never inscribed, and the rest, with the single exception of No. 345, are wholly illegible.

No. 345.

Albistan. In the cemetery. See Preliminary Report, *p. 37, No. 56. Copy.*

```
    C A E S
          E P
    ▨▨▨▨▨▨▨
          N ı ᴗ Λ
    ▨▨▨▨▨▨▨
    R E S T I T V T A
    P O F E L L I
    V M T H E O D O R V M
    A V G · P R  P R
```

[Imp.] Caes.
[divi Severi n]ep.,
[divi M. Antonini
       fil.
M. Aur. Antonino
Pio Felici Aug.
milia] restituta
[per M. Ul]p. Ofelli-
um Theodorum
[leg.] Aug. pr. pr.

Nos. 346–347.

See Bulletin de Correspondance Hellénique, 1883, *p.* 142, *No.* 30 : Sur une colonne, dans un champ, à un demi-mille anglais des deux inscriptions précédentes ; lettres très-

frustes. Copie de M. Ramsay. *See also* Ephemeris
Epigraphica, 1884, *p. 584, No.* 1366.

```
        C A I U A ⋔
      M A X I M I A N O
          O ⸀ I ⸀ C
             ⋖─
        C O N I A I ᴗ A
  6   R    T I T V E R V I T
      R ⸜ N T O N I V ⋋
        ⋔ ⋔ I V ⋔
        O N E ⋔ V C I E G
        A ⸜ G P R P R
 10        P N B
```

This inscription I did not see, as my line of march did not lie
along the valley of the Sarus above Comana.

The new light thrown upon the history of Cataonia by my milliaria
makes it certain that this inscription must be divided into two, the
restoration of both of which being beyond question. I venture to
insert it here mainly in order to clear up the doubts and questions
raised by Mr. Waddington in the *Bulletin* as cited above.

The original inscription (*A*), remnants of which are lines 4–10,
stood in the name of the Philippi Augusti. Mr. Waddington points
out that in case the inscription belongs to Diocletian and Maximian
Augusti and Constantius and Maximian Caesares, as line 2 would
seem to indicate, then the title *vir clarissimus legatus Augusti pro
praetore* is historically inaccurate, inasmuch as from the times of
Diocletian on the province was governed only by a *praeses* or *Con-
sularis*. But my Nos. 290, 292, 294, 310, etc., make it perfectly
clear that the stone held two inscriptions, the oldest of which, being
in the name of the Philippi, might well enough give the governor the
title *vir clarissimus legatus Augusti pro praetore*. From the same
inscriptions it is clear that Mr. Waddington's conjecture of *Senecio* as
the name of the legate is wrong, and that the name is Antonius Memmius Hiero.

The inscriptions read originally as follows :

A.

[Im-
(p). Caes. Ma-
rcus Iulius P-
hilippus Piu-
s Felix Invi-
ctus Aug.
et Marcus
Iulius Phi-
lippus no-
bilissimus
Caesar via-
s et pontes
vetustate]
con[l]a[ps]a[s]
r[es]titueru[n]t
p[er A]ntonium
[Me]mmium
[Hier]onem [l]eg.
A[u]g(ustorum) pr. pr.

B.

[Impp. Caess.
Diocletiano
et M. Aur. Val.
Maximiano
P. F. Invi. Augg.
et Flavi. Val.
Constantio
et G]a[l]. V[al]
Maximiano
[n]ob[b]. C[aess].

No. 348.

Bulletin de Correspondance Hellénique, 1883, *p.* 140, *No.* 27:
Sur une colonne, près de l'endroit où la voie romaine
devait entrer dans la ville.   Copies de MM. Clayton et
Ramsay.   *See also* Ephemeris Epigraphica, 1884, *p.* 36.
No. 75.

```
I////////C % V E
O %//A X I M I N O
% P I O % F E L I C I %
V I C T O % A V G
% P % M % T R I B
% P O T E S T % P P
```

I insert this here because the milestones found by me (see Nos.
[272], 293, 309) make its restoration certain.

[Imp. Caesari Gaio]
I[ulio] Ve[r-]
o [M]aximino
Pio Felici
[In]victo Aug.
p(ont). m(ax). trib.
potest. p(atri) p(atriae)
[per Licinnium
Serenianum
leg. Aug. pr. pr.]

The following inscription is inserted for the same reason.

No. 349.

See Bulletin de Correspondance Hellénique, 1883, *p.* 140,
No. 28: Dans une maison, à 5 milles anglais au N.E. de
Char.   Copie de M. Ramsay.   *See also* Ephemeris Epi-
graphica, 1884, *p.* 37, *No.* 76.

```
CAESG
IVLIO%\
MAXIM
PIO%FELIC
INVICTO
%P%M%T
POTEST
▨▨▨▨▨▨▨
PER
▨▨▨▨▨▨▨
LEGAVGPR
  MP    NГ
```

[Imp.]
Caes. G[aio]
Iulio [Vero]
Maxim[ino]
Pio Felic[i]
Invicto [Aug.]
p. m. t[rib].
potest.
p(atri) p(atriae)
per [Licinnium
Serenianum
leg. Aug. pr. pr.
  M PNГ

Owing to my severe illness we were detained three days in Albistan, during which time kind attentions were showered upon us by Rev. and Mrs. Henry Marden, American missionaries of Marash.

Henceforward no milliaria were found. I am wholly unable to account for this fact, as there are only two possible roads from Albistan to Melitene, one of which we traversed on the way out and the other on our return. It may be safely affirmed, however, that the Roman road did not go by way of Köz Agha and Pulat, since this whole road is much too difficult. Had the Roman road gone this way, it could not have avoided the abrupt pass of Ola Kaya, and it is exactly this pass that makes it necessary to look for it elsewhere. The only other route is that by way of·Derinde, and thence down the Tokhma Su to Malatia (the ancient Melitene).

August 8. Albistan to Yenikieui, 5 h. 29 m. There is a badly defaced Hittite inscription in the cemetery of Kütchük Yapalak. We traverse the great plain of Albistan. North of Böyük Yapalak we enter a narrow valley, which gradually ascends to the plateau on which Yenikieui is situated.

No. 350.

Ashagha Yapalak. In the cemetery. Letters very faint and · blurred. Copy.

H Δ E M E T □ N Δ Y

Δ I □ Δ □ T □ N Ξ H

Ⅽ H M A Δ A Δ ˉ Λ I I

T H N Δ E π π A I Ⅽ π □

August 9. Yenikieui, *via* Arslan Tash, to Köz Agha, 6 h. 12 m. We visited Arslan Tash and got photographs of the lions, discovered by von Moltke. They once stood on either side of a gateway just as the Assyrian Cherubim did. The Wolfe Expedition to Babylonia discovered similar lions at Arslan Tash in the Serudj Ova, a day's journey southeast of Biredjik in Mesopotamia. These Mesopotamian lions are of much better workmanship, and besides are better preserved. But the two pairs of lions belong, no doubt, to the same epoch.

The road southeast of Böyük Yapalak traverses an open rolling country; it is barren, for the most part, there being no means of irrigating it.

August 11. Köz Agha to Pulat, 9 h. 11 m. A journey of great difficulty, especially east of the Soghud Su, where the ascent to the pass of Ola Kaya Dagh begins. The country is very rough. The mountains are volcanic. The time from Köz Agha to the summit of the pass of Ola Kaya is 7 h. 35 m. The descent is very abrupt, and in places progress is almost impossible. In 1 h. 28 m. from the summit of the pass we reach the plain of Pulat.

No. 351.

Pulat. Stele by a fountain. Copy.

ΔΙΟΔΟΤΟΓΤΙΚΕΓΝΟΥ
ΗΛΙΑΔΙΟΥΑΓΟΥΤΗ
ΦΙΛΟΤΕΚΝѠΜΗΤΓΙ

Διόδοτος Τικέρνου
Ἡλιάδ⟨ι⟩ου Ἀρούτῃ
φιλοτέκνῳ μητρί.

August 12. Pulat to Kalaïk, 8 h. 19 m. Thirty-nine minutes north of Pulat we reach the low watershed, and thenceforth go down a small arm of the Sultan Ṭchai, which we cross a short distance east of Tchutlu. The eastern bank of the Sultan Tchai is a great bluff, which is ascended in 21 m. We then find ourselves on a great elevated plateau, which is broken by the two rivers west of Kalaïk. Kalaïk is situated on the western bluff of the river, and about six hundred feet above the river. A very large canal of ice-cold water flows through Kalaïk, and goes all the way to Malatia. It is this canal which furnishes the city of Malatia with its abundant supply of water. Besides this it irrigates the whole intervening country, which is a veritable garden spot. A great variety of fruit trees grow on every hand, and the fruit of Malatia is celebrated far and wide.

August 13. Kalaïk to Malatia, 1 h. 46 m. We pass through the delightful forest of fruit trees that extend all the way to Malatia. Their cool refreshing shade is delightful to the traveller after a journey of weeks through a treeless country. The new city of Malatia is reckoned as the half-way station on the overland route from Constantinople to Baghdad. It is a wide-awake business town, and in this respect it differs very materially from the ordinary Turkish town. When the Egyptians were at war with the Sultan a large number of Turkish troops were quartered for an indefinite period on the people of old Malatia, which stood on the site of Melitene. This was more than the long-suffering inhabitants could bear; so they abandoned their old homes to the soldiers, and built a new city among the gardens seven or eight miles southwest of Melitene. After the war-troubles were over the people still clung to their new abodes.

August 14. Malatia, *via* Melitene, to the junction of the Tokhma Su with the Euphrates, opposite Sheikh Hassan, 3 h. 16 m., and return to Malatia. Melitene is now a mass of ruins; among them many fine specimens of the ornamented architecture of the Seldjuks are conspicuous. The whole country between Melitene and the Euphrates is exceedingly fertile.

August 15. Malatia to Sara Hadji, 8 h. 37 m. West of Arga we cross a mountain to Kürdkieui; then comes a wild gorge and a steep ascent to Sara Hadji on the mountain side. Here our whole party escaped being murdered only by a miracle, and man and beast hungered until the night of the following day.

August 16. Sara Hadji to Müghde, 8 h. 45 m. Leaving Sara Hadji we reach the summit of the mountain in 43 m. Then we descend to another Kürdkieui, situated in a wild gorge, then another great mountain is crossed, and finally the Tokhma Su is reached at Bel-i-Gedik. At this point the river flows through narrows for half a mile. A perpendicular wall of rock, three or four hundred feet high, is on either side of the river. Consequently the road has to climb the little mountain. Once across this mountain we go up the open valley of the Tokhma Su to Müghde, where we halt a day to recruit ourselves and horses after our long fast. The whole mountain country between Arga and the Tokhma Su is inhabited solely by Kurds, an inhospitable, murderous set of filthy villains, who still preserve all the ferocious characteristics of their ancestors, the ancient Καρδοῦχοι, of whom Xenophon has little good to report in the *Anabasis*.

August 18. Müghde, *via* Derinde, to Yenikieui, 6 h. 22 m. The valley between Müghde and Derinde is very fertile. An hour east of Old Derinde the valley contracts to a gorge, and New Derinde stretches out on both sides of the river for the whole distance between this point and the Derinde of the old map. Old Derinde was abandoned like Old Malatia, and for the same reason. It is now a grand mass of ruins. Derinde means "*in* or *at* the gorge." Professor Kiepert regards the name as a popular interpretation of the ancient name Δελενδίς. At Old Derinde the river has cut its way through the solid rock, which rises perpendicularly to a height of three or four hundred feet on either side of the river. The width of the pass through which the river thus flows is about fifty feet. On the right bank is the almost impregnable castle, probably dating from the time of the early Turks; at the foot of the castle and west of it lies the abandoned town.

  Ashta is also situated in a gorge. The top of the eastern bluff corresponds with the general level of the surrounding plateau. Up to this point the country is difficult and our progress slow. When 2 h. 13 m. out from Derinde we found a small lion in black basalt by the roadside. Photographs were taken, but it was just growing dark, and they did not succeed well. After leaving the lion a heavy thunderstorm overtook us ; my men got separated into four parties, each of which got lost. It was about midnight when we were all together again at Yenikieui. It would have been a sad night for some of us, but for the generous exertions of our good Zaptieh Halil.

  August 19. Yenikieui to Böyük Tatlar, 6 h. 49 m. We cross a mountain between Yenikieui and Ketchi-Maghara. Thenceforward the country is open and rolling.

  August 20. Böyük Tatlar to Örtülü, 6 h. 14 m. The country between Böyük Tatlar and Kereïkieui is very rough and mountainous. From Kereïkieui we go down the gorge of the Khurman Su to Khurman Kalesi. Khurman Kalesi is a proud castle, possibly of early Turkish origin, situated on a crag just at the junction of the Maragos Tchai with the Khurman Su.

Nos. 352–354.

On the living rock twenty-three minutes northwest of Khurman Kalesi. See my Preliminary Report, *p.* 39, *Nos.* 57, 58, 59. *Copy. Photographs of B and C.*

  Inscription *A* consists of eight heroic hexameters ; *B*, of two hexameters ; and *C* is an elegiac distich. *B* and *C* cannot be reached without artificial help, which everywhere in Turkey it is difficult to obtain. Of these two we got photographs. Inscription *A can* be reached, but only with danger to life or limb. The letters are immense, and the surface covered by the inscription is so great that only a few letters in each line can be read at a time ; this done, one must climb down and then up again, it being impossible to move horizontally along the face of the rock. Copying the inscription was very laborious work, as I had to remove my shoes and support myself by my toes. First the moss was removed from the letters, then the inscription was copied, and lastly the copy was verified.

A.

ΑΚΙΛΛΙΟΥΧΕΙΡΙCΟΦΟΥΑΛΕΞΑΝ
ΔΡΟΥΤΟΥΚΑΙΦΙΛΙΠΠΙΟΥ
ΤΗCΔΕΚΟΡΗCΚΟΠΗCΠΟΤΑΠΗΛΙΒΑΤΟΙΟΘΟΡΟΥCΑ
ΑΘΑΝΑΤΩΝΒΟΥΛΗΙCΙΝΥΤΕΚΦΥΓΕΝΑΡΚΤΟΝΑΤΗΜΩΝ
5 ΔΙΧΘΑΔΙΗΙCΚΩΜΗΙCΙΦΙΛΙΠΠΙΟΥΑΡCΙΝΟΟΥΤΕ
ΟΥΤΟCΑΡΙΓΝΩΤΟCΠΡΕΙΩΝΟΡΟCΑCΤΥΦΕΛΙΚΤΟC
ΕΠΛΕΤΟ·Δ'ΑΡCΙΝΟΩΙΜΕΝΕΔΕΘΛΙΑCΑΡΡΟΜΑΗΝΑ
ΤΩΙΔ'ΑΡΕΠΙΠΡΟΧΟΗΙCΙΔΥΩΤΟΤΑΜΩΝCΟΒΑΓΗΝΑ
ΠΙCΤΟΙΔ'ΑΛΛΗΛΟΙCΕΤΑΡΟΙΠΕΛΟΝΩΝΦΙΛΟΤΗΤΑ
10 ΑΡΡΗΚΤΗΝΠΑΓΟCΟΥΤΟCΑΠΑΓΓΕΛΛΟΙΚΑΙΕΠΕΙΤΑ

B.

ΤΟΥΑΥΤΟΥΧΕΙΡΙCΟΦΟΥ
ΕΝΝΕΑΤΟΙΠΕΤΡΗΘΕΕΝΕΠΙΚΡΗΝΗΝCΟΒΑΓΗΝΩΝ
ΚΑΛΛΙΡΟΟΝCΤΑΔΙΟΙΚΟΡΑΚΟCΠΟΤΑΜΟΙΟΠΑΡΟΧΘΑC

C.

ΤΟΥΑΥΤΟΥΧΕΙΡΙCΟΦΟΥ
ΕΓΓΥΘΙΤΟΙCΟΒΑΓΗΝΑΚΑΙΑΙΓΛΗΕΝΤΑΛΟΕΤΡΑ
ΗΝΔΟΛΙΓΟΝCΠΕΥΧΗΙCΛΟΥCΕΔΕ̣ΚΚΑΜΑΤΟΥ

A.

'Ακιλλίου Χειρισόφου 'Αλεξάν-
δρου τοῦ καὶ Φιλιππίου.
Τῆσδε κόρη σκοπιῆς ποτ' ἀπ' ἠλιβάτοιο θοροῦσα
ἀθανάτων βουλῇσιν ὑπέκφυγεν ἄρκτον ἀπήμων ·
5 διχθαδίης κώμῃσι Φιλιππίου 'Αρσινόου τε
οὗτος ἀρίγνωτος Πρείων ὄρος ἀστυφέλικτος.
ἔπλετο δ' 'Αρσινόῳ μὲν ἐδέθλια Σαρρομάηνα,
τῷ δ' ἄρ' ἐπὶ προχοῇσι δύω ποταμῶν Σοβάγηνα
πιστοὶ δ' ἀλλήλοις ἔταροι πέλον, ὧν φιλότητα
10 ἀρρήκτην πάγος οὗτος ἀπαγγέλλοι καὶ ἔπειτα.

B.

τοῦ αὐτοῦ Χειρισόφου.
'Εννέα τοι πέτρηθεν ἐπὶ κρήνην Σοβαγήνων
καλλίροον στάδιοι Κόρακος ποταμοῖο παρ' ὄχθας.

C.

τοῦ αὐτοῦ Χειρισόφου.
Ἐγγυθί τοι Σοβάγηνα καὶ αἰγλήεντα λοετρά ·
ἢν δ' ὀλίγον σπεύσῃς [λ]ούσ[εα]ι ἐκ καμάτου.

These inscriptions may be translated as follows :

A.

" Epigram of Acilius Chirisophus, the son of Alexander ; also
called Philippius. ,
Once upon a time, by the counsels of the Immortals, a girl rushed
down from this lofty crag and escaped unhurt from a bear. To the
two villages of Philippius and Arsinous this well-known Prion is a
boundary not to be disturbed. The home of Arsinous was Sarromaëna ; that of Philippius was Sobagena, at the confluence of two
rivers. They were faithful comrades, and may this rock declare their
unbroken friendship even to future ages."

B.

" Of the same Chirisophus.

It is nine stadia from this rock to the fair-flowing spring of Soba-gena, on the bank of the river Korax."

C.

" Of the same Chirisophus.

Near by is Sobagena with its bright clear baths. If you will hasten a little, you may bathe yourself after your toil."

Line 3 of *A*. By consulting my *Preliminary Report* on this jour-ney, p. 39, it will be seen that I inserted in the uncial text the letter C in brackets, thus [C], meaning to indicate thereby that this C was not on the rock. At first it was my intention to give only the uncial text of the inscriptions, as I had done throughout the *Report*, but as I attached importance to them, I afterwards inserted the minuscule text as well. I then forgot to erase the [C] of the uncial text.

The readings of Professor Allen (*Preliminary Report*, p. 41, top) are all undoubtedly correct except [ρ]ώμῃσι and ὅρος, which are clearly wrong. The inscriptions are of prime importance for the topography of this region.

A girl, when pursued by a bear, had rushed down over the almost perpendicular crag, which raises its proud head to a height of about 1500 feet. By a veritable miracle she escaped unhurt, and as a lasting memento of this great escape, two friends, Philippius and Arsinous, one possibly her father, had these inscriptions engraved on the rock. From the inscriptions it is clear: 1°, that Khurman Kalesi occupies the site of Sobagena, the village of Philippius, inasmuch as it is situated at the junction of the Maragos Tchai with the Khurman Su; 2°, it is clear that the ancient name of Khurman Su was the Korax; 3°, it is clear that Sarromaëna, the village of Arsinous, must have occupied the site of Maragos, which name may even be a cor-ruption of Sarromaëna; 4°, it is clear that the mountain bore the name of Prion. Thus from these inscriptions we locate and give names to two villages, a river, and a mountain. The rock on which *A* is inscribed is certainly " a boundary not to be disturbed," and nine stadia is about the true distance (23 m.) from Khurman Kalesi to the rock which bears inscription *B*.

Leaving these inscriptions, we go up the little valley of the Maragos Tchai past Maragos to Topak Tash (not Toprak Tash). Here we leave the gorge, and ascend a great mountain to an elevated plateau inhabited by inhospitable Kurds.

August 21. Örtülü to Savoghlan, 9 h. 34 m. We descend by a rough mountainous road to the valley of the Seihûn. The valley is of respectable size here, and contains a number of villages. The name of the district is Saris. We were just 1 h. 16 m. in crossing the valley from mountain to mountain. In crossing the mountain west of the valley of the Seihûn 1 h. 15 m. are taken up. Thenceforward we go down a narrow valley until the open country is reached in the neighborhood of Bagtchekieui.

August 22. Savoghlan to Seresek, 6 h. 0 m. Fifty minutes west of Savoghlan we ford the Zamantia Tchai in the neighborhood of Kizilkhan. Ekrek is probably the site of an ancient town. At Karadaghi there is a good Seldjuk Khan. Seresek is the ancient Arasaxa.

August 23. Seresek to Talas, 5 h. 27 m.

August 25. Talas, *via* Kaisariye, to Indjesu, 6 h. 17 m.

August 26. Indjesu to Ürgüp, 5 h. 9 m. We travelled all day in a rain, so that our progress was slow. Leaving Indjesu, we cross a ridge, and in 2 h. 37 m. we are down at Akkieui, at the head of the very fertile valley that leads hence to Ürgüp. Ürgüp is a prosperous town, well built of the soft volcanic tufa. The whole region of country between Ürgüp and Tatlar is in reality an extinct volcano.

August 27. Ürgüp to Udjessar, 2 h. 16 m. Martchan is the centre of the cone formations and of the rock-cut dwellings. The scenery is wonderful beyond all description. We spent this day and the most of August 28th in securing a large number of photographs of the cones and rock dwellings. The character of the rock-cut dwellings of Martchan and Udjessar is the same as that of those at Selme and Soghanlü Dere, already described above. Only here they are more abundant, and the volcanic character of the country is much more marked.

August 28. Udjessar to Nevshehir, 1 h. 10 m. Nevshehir is a large and prosperous town, with excellent mosques and theological schools.

August 29. Nevshehir to Tatlar, 2 h. 49 m. The country between the two places is one vast and barren lava-field.

August 30. Tatlar to Hadji Bektash, 5 h. 53 m. The Halys is wide, but not deep, at the point where we forded it. North of Salanda we crossed a spur of Khirka Dagh to the great plain of Hadji Bektash, the headquarters of the Dervishes and the tomb of Hadji Bektash himself. We were entertained with distinction by the Dervishes. There are great salt-mines in the neighborhood.

September 1. Hadji Bektash to Karaseñir, 7 h. 29 m. It was my purpose to explore the unknown region between Hadji Bektash as well as could be done on a straight march. The results are laid down in the map of Northern Cappadocia which accompanies this volume.

The country northeast of Hadji Bektash, as far as Tchroprun Oghlu is mostly level. Here we go down the gorge of a little river to Doiduk, then cross a ridge to Kazaklü, from which point we traverse a plain to Karaseñir.

September 2. Karaseñir to Hadji Shefa'atli, 5 h. 27 m. Between Karaseñir and Kediler the country is undulating; at Kediler the plain of Pashakieui is entered.

September 3. Hadji Shefa'atli to Yerkieui, 5 h. 48 m. Fourteen minutes east of Hadji Shefa'atli is the junction of the Kara Su with the Kanak Su; henceforward the united stream is called the Delidje Irmak. At this point it enters a cañon, which continues as far as Öyük, where it enters the plain. This cañon is so abrupt and precipitous that the road cannot follow it, but ascends to an elevated plateau, on which are the villages Djafali and Adjikoyun. From this point there·is a gradual descent to the cañon, which is still impassable, and the road crosses a series of ridges on the right bank of the river to Öyük.

September 4. Yerkieui to Böyük Nefezkieui, 4 h. 42 m. We travelled very rapidly from Boyalik to Böyük Nefezkieui.

No. 355.   •

Boyalik (called also Medjidic). Panel in a slab.   Copy.

ΕΝΘΑΔΕΚΑ
ΤΑΚΙΤΕΟ
ΜΑΚΑΡΙΟC
ΓΥΜΝΑCΙC
ΚΥΡΙΕΕΛΕΗ
CΟΝΤΟΝΔΟΥΛΟ
ΝCΟΥ

'Ενθάδε κα-
τακῖτε ὁ
μακάριος
Γυμνάσις.
Κύριε ἐλέη-
σον τὸν δοῦλό·
ν σου.

At Kütchük Nefezkieui there is a large spring, which is the chief source of the stream up which we came from Khatibinkieui. The spring was formerly used as Baths, the ruined walls of which still stand. At Kötlak there are many architectural remains, but no inscriptions, so far as I could discover.

No. 356.

Böyük Nefezkieui (Tavium). Roman milliarium in a cemetery between Böyük Nefezkieui and Assara, and immediately west of the Acropolis of Böyük Nefezkieui. See Preliminary Report, p. 43, No. 60. Copy.

```
              I M P
N E R V A C A E S A R A V
P O N T M A X T R I B P O T E S V I I
C O S I I I P P R E S T I T V I T
P E R P O M P O N░░░M
B A S S V M L E G P R O P R
```

P M I̅ A

Imp(erator)

Nerva Caesar Au[g](ustus)

Pont(ifex) Max(imus) trib(uniciae) potes[t](atis) II

Co(n)s(ul) III, p(ater) p(atriae) restituit

per Pompon[iu]m

Bassum leg(atum) pro pr(aetore).

P(assuum) m(ille) I, a′.

The number of miles is given in both Latin and Greek, as seems to be indicated by the horizontal bar over the I̅. This becomes certain when the milestone found by Professor Hirschfeld at Iskelib [see Hirschfeld's article *Tavium* in the *Sitzungsberichte der königl. preuss. Akademie der Wissenschaften zu Berlin*, 1883, Vol. LIII. p. 1256, and *Ephemeris Epigraphica*, 1884, p. 39, No. 81] is compared with the inscription given above. It, too, records a repair of roads by this same legate Pomponius Bassus and its numerals

$$\text{M I L} \cdot \text{P} \cdot \text{L} \overline{\text{X X X}}$$
$$\overline{\overline{\pi}}$$

are certainly bilingual.

Two other inscriptions of this legate are known [see *C.I.L.* III. 309, and *Journal of Philology*, 1882, p. 155 = *Ephemeris Epigraphica*, 1884, p. 39, No. 82].

For a discussion of the date when T. Pomponius Bassus governed Galatia, Cappadocia, Pontus, etc., see *Journal of Philology*, 1882, pp. 155, 156; *Bullettino dell' Instituto*, 1844, p. 125 sqq., 1862, pp. 67, 68; *Annali dell' Instituto*, 1844, pp. 14 and 40; Eckhel, *Doct. Num.*, III. p. 190; Mionnet, *Suppl.* 7, pp. 632, 665, 669; Perrot,

de Galat. prov. Rom., p. 111. Pomponius Bassus is mentioned as
πρεσβευτής in an inscription of Ephesus recently published in the
Mittheilungen des Deutschen Archaeologischen Institutes in Athen,
1885, p. 401.

The above inscription (No. 356) is one of the most important
discoveries of the journey. The ancient Tavium was the ἐμπόριον
τῶν ταύτῃ. It was of prime importance geographically, because it
was the centre from which diverged seven roads, five of which are
given in the Peutinger Table, and the remaining two in the Antonine
Itinerary. Distances along these roads were measured from Tavium ;
consequently it was of the highest importance to discover the real
site of Tavium, for on it depends the geography of the whole country
between Ancyra and Amasia. Tavium has been located by different
scholars at Tchorum, Böyük Nefezkieui, Boghazkieui ; but until
recently those best entitled to an opinion had settled on Böyük
Nefezkieui as the true site, but always, be it understood, without any
documentary proof. In November, 1883, Professor Gustav Hirschfeld,
of Königsberg, published an article "*Tavium*" in the *Sitzungsberichte
der Academie der Wissenschaften zu Berlin*, in which he declined to
accept for Tavium any of the sites hitherto suggested. He attempts
to show that Tavium must be sought on the left bank of the Halys,
and that its site is occupied by *Iskelib*, a degree north of Böyük
Nefezkieui. In January, 1884, Professor Heinrich Kiepert published
in the *Sitzungsberichte* (as above) his *Gegenbemerkungen zu der
Abhandlung des Hrn. G. Hirschfeld über die Lage von Tavium*, from
which it appears that he is very loath to give up the site of Böyük
Nefezkieui as that of Tavium ; but he finally suggests Aladja, or a
point immediately southeast of Aladja.

Now my inscription (No. 356) is the first milestone from somewhere, and as distances in this region were reckoned from Tavium, it
necessarily follows that it is the first milestone on the Roman road
from Tavium to Ancyra, and consequently Tavium is located beyond
dispute at Böyük Nefezkieui. But to make the matter doubly sure
there is still another point to be taken into consideration. In the
cemetery of Tamba Hassan, a village just two hours north of Böyük
Nefezkieui, Mr. Haynes found Roman milliaria, one of which bore
the badly defaced inscription No. 377. Now, as I understand it,
Tamba Hassan is none other than the *Tomba* or Tonea of the

Peutinger Table, the first station on the Roman road from Tavium to Comana in Pontus. Hirschfeld points out that Tomba and Tonea are two names for the same place. It must be noted that the distances, as given by the Peutinger Table, viz. Tonea XIII and Tomba XVI MP. from Tavium, do not agree accurately with my identification, and I should rather look for VIII instead of either XIII or XVI. The Table is almost certainly in error, and the identification both of Tavium and Tomba remains fixed.

It has been stated that the ruins of Böyük Nefezkieui are too insignificant to represent Tavium. This is not the case. It is true that at the village itself there are only comparatively small fragments; but the cemeteries, both of Kötlak and the one in which No. 356 was found, are full of architectural fragments, and the last-mentioned cemetery has scarcely any other stones in it except cippi, columns, and fragments of epistyles, all of considerable weight and size. A future traveller will no doubt find the hot springs in the region of country between Böyük Nefezkieui and Yozgad.

I found only Roman coins at Böyük Nefezkieui, of the Caesarean coinage. The soil is very fertile, and yields abundant harvests of wheat; and the people plant nothing else.

No. 357.

Böyük Nefezkieui. Ornamented epistyle of white marble.
See Bulletin de Correspondance Hellénique, 1883, *p.* 26, *whence it was inserted in the* Ephemeris Epigraphica, 1884, *p.* 28, *No.* 42. *Copy.*

▓▓▓P E R A T O R V⫯ C O▓▓▓
▓▓▓Γ

[Im]perator VI Co[(n)s(ul)].

No. 358.

Böyük Nefezkieui. Stele in the wall of a house. Copy.[1]

---

[1] Ligatures occur: line 3, WN, MH; line 4, MH.

▨▨▨▨ΥΦΙΝΑΑΣΚΛΗ

▨▨▨▨ΔΗΣΥΝΒΙΩΛΙ

▨▨▨▨ΥΡΓΩΝΙΚΟΜΗΔΙ

▨▨▨▨ΜΗ   ΣΧΑ

[wreath]   ΡΙΝ

['Ρο]υφῶνα 'Ασκλη-
[πιά]δη? συνβίῳ Λι-
[κο]ύργῳ Νικομηδί-
[ου? μνή]μης χάριν.

From the following inscriptions it is clear that Tavium was a stronghold of Christianity.

No. 359.

Böyük Nefezkieui. Black stone. Copy.

ΕΝΘΑΚΑΤΑ
ΚΙΤΕΗΔΟΥ
ΛΗΤΟΥΧ̄Υ
ΤΟΥΑΛΥΠΙΑ

Ἔνθα κατα-
κῖτε ἡ δού-
λη τοῦ Χ(ρισ-)
τοῦ 'Αλυπία.

No. 360.

Böyük Nefezkieui. Copy.

ΤΟΥΘ̄Υ
ΘΕΟₚШ
ΡΟϹΟ
ΖΟΥΒΛΟ
Ϲ†

[Ἔνθα κατα-
κῖτε ὁ
δοῦλος]
τοῦ θ(εο)ῦ
Θεόδω-
ρος ὁ
Ζοῦβλος.

No. 361.

Böyük Nefezkieui. Copy.

Є Ν Θ Α Κ Α Τ Α
Κ Ι Τ Є Η Δ Ο Υ Λ Η
Τ Ο Υ Θ͞Υ
Π Є Λ Α Γ Ι Α

Ἔνθα κατα-
κῖτε ἡ δούλη
τοῦ θ(εο)ῦ
Πελαγία.

No. 362.

Böyük Nefezkieui. Copy.

✝ Є Ν Θ Α Κ Α Τ Α
Κ Ι Τ Є Η Δ Ο Υ Λ▨
Τ Ο Υ Θ Є Ο Υ
Θ Є Ш Δ Ο Τ▨
✝

Ἔνθα κατα-
κῖτε ἡ δούλ[η]
τοῦ θεοῦ
Θεωδότ[η].

No. 363.

Böyük Nefezkicui.   Copy.

+ K Y M H Є
A Λ Y Π Ι A C
Δ O Y H C X̄ Ȳ
✝

Κύμησ(ις)
'Αλυπίας
δού(λ)ης Χ(ριστο)ῦ.

Κύμησις stands for κοίμησις.

No. 364.

Böyük Nefezkicui.   Copy.

Є N Θ A K A
T A K Ι T Є
O Δ O Y Λ O C
T O Y Θ̄Ȳ
Γ Є O P Γ Ι C

*Ἔνθα κα-
τακῖτε
ὁ δοῦλος
τοῦ θ(εο)ῦ
Γεόργις.

No. 365.

Böyük Nefezkicui.   Copy.

▨▨▨▨Θ A
K A T A

```
K I T E O Δ
Ʀ Λ O C
T Ʀ O̅Y̅
Δ A N I
H Λ
```

⳨

["Εν]θα

κατα-

κῖτε ὁ δ-

οῦλος

τοῦ θ(εο)ῦ

Δανι-

ήλ.

No. 366.

Böyük Nefezkieni.   Copy.

```
E N Θ A K A
T A K I T E
O Δ O Y Λ O C
T O Y Θ Y Π ΛΤ
Λ O C Π P O
T O Π P E C
B Y T E P O C
```

῎Ενθα κα-

τακῖτε

ὁ δοῦλος

τοῦ θ(εο)ῦ Π[αῦ-]

λος προ-

τοπρεσ-

βύτερος.

No. 367.

Böyük Nefezkieni.   *Copy.*

```
† Є N Θ A
  K A T A K
  I T Є O Δ ४
  Λ O C T ४
    Θ̄ Ȳ Γ
  Є P M ⸏
  ◌ ◌
```

Ἔνθα
κατακ-
ῖτε ὁ δοῦ-
λος τοῦ
θ(εο)ῦ Γ-
ερμ[α-]
[νοῦ].

No. 368.

Böyük Nefezkieni.   *Copy.*

```
Є N Θ A
K A T A
K I T Є
O Δ O Y
Λ O C
T Ω Y Θ̄ Ȳ
C T C
Φ A N O C
```

Ἔνθα
κατα-
κῖτε

ὁ δοῦ-
λος
τοῦ θ(εο)ῦ
Στ[έ]
φανος.

No. 369.

Böyük Nefezkieni. Copy.

```
    E N
 □ A 'K Λ
 T A K I
 T E H Δ □ Y
 Λ H T □ Y
 □Y ⊏ T E
 Φ A N I Ϲ
```

Ἐν-
[θ]α κα-
τακῖ-
τε ἡ δού-
λη τοῦ
[θ](εο)ῦ Στε-
φανίς.

No. 370.

Böyük Nefezkieni. Copy.

```
▨N Θ A K A
▨A K I T E O
▨Ȣ Λ O C T Ȣ
▨E Ȣ E Y Ⱥ
▨M I C
          †
```

['Ε]νθα κα-
[τ]ακῖτε ὁ
[δ]οῦλος τοῦ
[θ]εοῦ Εὐ[δ-]
[ά]μις.

No. 371.

Böyük Nefezkieui.   Copied by J. H. Haynes.

ΕΝΘΑ
ΚΑΤΑΚΙΤΕ
ΗΔΟΥΛ
Η
ꞶΟΥΘΕΟ
ΥΙꞶΑΝΝ
ΙΑ

Ἔνθα
κατακῖτε
ἡ δούλ-
η
[τ]οῦ θεο-
ῦ Ἰωανν-
ία.

The form Ἰωαννία is a new form of the name.

No. 372.

Böyük Nefezkieui.   Copied by J. H. Haynes.

✝ΕΝΘΑΚΑ
ΤΑΚΙΤΕΟ
ΛϹΥΛΟϹ

Ἔνθα κα-
τακῖτε ὁ
[δοῦ]λος
[τοῦ θ(εο)ῦ]

. . . .

No. 373.

Böyük Nefezkieui. Copied by J. H. Haynes.

E N
K A
K I T
O Y Λ
H A
A

Ἔν[θα]
κα[τα-]
κῖτ[ε]
[ἡ δ]ουλ-
η κ.τ.λ.

No. 374.

Böyük Nefezkieui. Black stone in the wall of a house. Copy.[1]

ΓΑΡΓΗΤΙ   ΤΓΑΛΝ
        Γ Γ   ΓΛΔ
ΛΑΓΑΓΛ   ΙΓΑΙΟΓΥ
ΟΓΜΝΗ   ΛΑΡΙΝ

---

[1] Ligatures occur: line 1, HTT; line 4, NH.

No. 375.

Böyük Nefezkieui.   Epistyle block of white marble.   Copy.

▨Σ Ο Φ Ο Υ Α Π Ο Μ Ο Υ Σ Ε Ι Ο Υ▨

No. 376.

Böyük Nefezkicui.   Copy.

P I C

September 5.   Böyük Nefezkieui to Boghazkieui, 4 h. 52 m.   The road traverses a mountainous country.   We got photographs of the ancient and well-known rock sculptures.

No. 377.

Tamba Hassan.   The stone is partly embedded in the ground in the cemetery.   Copied by J. H. Haynes.

I S S I

R

I A

E N E O

O

C P O T

O

September 6.   Boghazieui to Öyük, 4 h. 34 m.   The ancient sculptures were photographed.

September 8.   Öyük to Ashagha Beshbunar, 5 h. 46 m.   The results henceforward were purely chorographic, and have been laid down in the map of Northern Cappadocia.

September 9.   Ashagha Beshbunar to Ulaklü, 8 h. 20 m.   The plain of Sungurlu comes to an end immediately west of Aghabunar, from which place the country is hilly to the Delidje Irmak. At Taobas we ascend a large mountain, which turns out to be the bluff of a great elevated plateau which extends from this point westward to the Kizil Irmak (Halys).

September 10. Ulaklü to Yalüm, 7 h. 14 m. In 6 h. 14 m. we reach the bridge over the Halys. The gorge through which the river flows abounds in vineyards, the ripe fruit of which was being converted into raisins. Hence a rough ascent of one hour to Yalüm.

September 11. Yalüm to Arablar, 9 h. 51 m. We traverse a rough country for 5 h. 41 m., when we reach the wagon road from Angora to Kaledjik.

No. 378.

Ortakieui. In the cemetery. See Preliminary Report, *p.* 45, *No.* 62. *Copy.*

```
IMPCAESVAI
SEVERO
℞OPIOFEL·IN
VICTOAVG·TRIB
5 POTIICOSI
```

M P

Imp(eratori) Caes(ari) [M]. A[ur](elio)
Severo [Alexand-]
[r]o Pio Fel(ici) In-]
victo Aug(usto) trib(uniciae)
5 pot(estatis) II Co(n)s(ul) I
M(ilia) P(assuum)? or else μ(ίλια) ρ'.

A mate to this inscription, found at Tchañly Kaya, an hour south of Ancyra, is *C.I.L.* III. 316. The date of both is 223 A.D.

If the reading of line 6 be μίλια ρ', then this is the one hundredth milestone from Tavium on the road to Ancyra.

September 12. Arablar to Angora, o h. 58 m. At Angora we were compelled to consider our journey finished, scientifically speaking. It was necessary for Mr. Haynes to reach Nicomedia by a certain day, in order to take the evening train for Constantinople, and our one thought thenceforth was to travel westward as rapidly as possible. For the sake of completeness I give the time from Angora to the railway terminus at Ismid.

September 13. Angora to Ayash, 7 h. 34 m.
September 14. Ayash to Kavun Ovasü Tchiftlik, 8 h. 59 m.
September 15. Kavun Ovasü Tchiftlik to Nali Khan, 7 h. 16 m.
September 16. Nali Khan to Köstebek, 5 h. 54 m.
September 17. Köstebek to Torbalü, 8 h. 40 m.
September 18. Torbalü to a Khan, 8 h. 31 m.
September 19. Khan to Sabandja, 9 h. 43 m.
September 20. Sabandja to Ismid, 5 h. 33 m.

ADDITIONAL NOTES.

—

No. 12 has been published by Ramsay in the *American Journal of Archaeology*, 1888, p. 346.

No. 21. Those who may be interested in "descent reckoned μητρόθεν" will find a treatment of the subject in Treuber's *Geschichte der Lykier*, p. 117 sqq.

Page 26, May 31. I have ascribed the identification of Apollonia to Paris and Holleaux, but Waddington located Apollonia at Medet before them.

No. 32. Published by Smith in the *Journal of Hellenic Studies*, 1887. The *Journal* is inaccessible to me, and I have not seen Smith's article.

No. 34. Published by Smith in the *Journal of Hellenic Studies*, 1887.

No. 35. Published by Ramsay in the *American Journal of Archaeology*, 1887, p. 365. In his uncial text, lines 10–12, he reads:

ΓΑΛΩΝΙΑ
ΓΕΒΑΓΤΗΙΙΗΤΓ▨ΛΚΑ
ΛΙ ΟΙ ΙΒΥΙ

and transcribes lines 9–12 as follows:

Σεπ-

τιμίῳ Γέτᾳ υἱῷ? τῶν
με]γάλων [Β]α[σιλέων καὶ ᾿Ιουλίᾳ
Σεβαστῇ μητέρα (sic) Κά[στρων.
᾿Α[π]ὸ [Κ]ιβύ[ρας Μιλια δυώδεκα?

No. 36. Published by Smith in the *Journal of Hellenic Studies*, 1887.

No. 37. Published by Ramsay in the *American Journal of Archaeology*, 1887, p. 363. In line 1 he reads A; in line 3, end, NA; in line 6, init., KAI.

Page 38, June 6. Ramsay says that Yimru Tash is the true name. I cannot agree with him.

Nos. 43, 44 (46). Ramsay writes that he verified the numerals of these inscriptions in 1886, and thinks that the date is PΠB.

Nos. 56–58. Published by Smith in the *Journal of Hellenic Studies*, 1887.

Nos. 62–63. Published by Smith in the *Journal of Hellenic Studies*, 1887.

No. 64. Ramsay (in the *American Journal of Archaeology*, 1887, p. 363) identifies Θεὸς Σώζων with Σαβάζιος. The god Σώζων is named on coins of Antiochia ad Maeandrum also.

It should be noted that if the era be the Asiatic, then the date of the inscription is either (TKZ) 243 or (TZ) 223 A.D.

No. 65. Published by Ramsay in the *American Journal of Archaeology*, 1887, p. 362.

If the era be the Asiatic, then the date is 168 A.D.

Nos. 72–75. Published by Smith in the *Journal of Hellenic Studies*, 1887. I have not seen his article.

No. 82. Published by Smith in the *Journal of Hellenic Studies*, 1887.

No. 83. Published by Smith in the *Journal of Hellenic Studies*, 1887.

No. 84. Ramsay now identifies the ancient site near *Einesh* with "Tymbrianasa, the modern name being the second half of the ancient name."

ADDENDA.

No. 379.

Tralleis. Forwarded to me in February, 1886, *by M. Mich. Pappa Konstantinou.* "On the base or capital of a column." *See* Mittheilungen d. K. Deutsch. Arch. Instituts, Athenische Abtheil., 1886, *p.* 203, *and* Bulletin de Correspondance Hellénique, 1886, *p.* 456.[1]

ΙΙΟΥΛΙΟΝ-Ι-ΙΟΥΛΙΟΥΦΙΛΙΠΠΟΥΑΡΧΙΕΡΕΩΣ
ΑΣΙΑΣΥΙΟΝΟΥΕΛΙΝΑΦΙΛΙΠΠΟΝΙΠΠΕΑΡΩ
ΜΑΙΩΝΤΩΝΕΚΛΕΚΤΩΝΕΝΡΩΜ··ΙΔΙΚΑΣΤΩΝ
ΕΠΙΤΡΟΠΟΝΤΩΝΣΕΒΑΣΤΩΝΠΑΤΕΡΑΙΟΥΛΙ·
5 ΦΙΛΙΠΠΟΥΣΥΓΚΛΗΤΙΚΟΥΣΤΡΑΤΗΓΟΥΡΩΜΑΙ
ΩΝΙΕΡΕΑΔΙΑΒΙΟΥΤΟΥΔΙΟΣΤΟΥΛΑΡΑΣΙΟΥ

[Γ.] Ἰούλιον, [Γ.] Ἰουλίου Φιλίππου ἀρχιερέως
Ἀσίας υἱὸν, Οὐελίᾳ, Φίλιππον, ἱππέα Ῥω-
μαίων τῶν ἐκλεκτῶν ἐν Ῥώμ[η]ι δικαστῶν
ἐπίτροπον τῶν Σεβαστῶν, πατέρα Ἰουλί[ου]
5 Φιλίππου συγκλητικοῦ στρατηγοῦ Ῥωμαί-
ων ἱερέα διὰ βίου τοῦ Διὸς τοῦ Λαρασίου.

See the note to the following inscription.

Var. Lect.

Mittheilungen reads: line 1, ΠΟΥΛΙΟΝ ΙΟΥΛΙΟΥΦΑΝΥΙΟΥ; line 3, ΜΑΙΟΝ and ΡΩΜΗΔ; line 4, end, ΙΟΥΛ.

---

[1] Ligatures occur in lines 2, ΠΠΕ; 3, ΝΕ bis.

No. 380.

Tralleis. Forwarded to me in April, 1886, by Mr. Mich. Pappa Konstantinou. It was found in the house of de la Chapelle, and published by Mr. Pappa Konstantinou in the Ἀμάλθεια *of Smyrna, April 27–29, 1884.*

```
ΙΟΥΛΙΟΝΦΙΛΙΠΠΟΝ
ΕΠΙΤΡΟΠΟΝΤΟΝΣΕΒΑΣ
ΤΟΝΠΑΤΕΡΑΙΟΥΛΙΟΥ
```

[Γ.] Ἰούλιον Φίλιππον
ἐπίτροπον τ[ῶ]ν Σεβασ-
τ[ῶ]ν, πατέρα Ἰουλίου
[Φιλίππου συγκλητικοῦ κτλ. See last No.].

Concerning C. Iulius Philippus, see the exhaustive study of Light-foot, *Apostolic Fathers*, Part II. Vol. I. pp. 612–618, and *Papers of the American School at Athens*, Vol. I. pp. 100–104.

From these two inscriptions (Nos. 5 and 6) it is clear that another member must be added to this distinguished family, and that the family tree is as follows:

C. Iulius Philippus, Chief-priest and Asiarch.
  |
C. Iulius Philippus, Roman Knight and *procurator Augustorum*.
  |
Iulius Philippus, Roman Senator and Priest of Zeus Larasios.

No. 381.

Tralleis. Forwarded to me in April, 1886, by Mr. Mich. Pappa Konstantinou. "In the house of Hadji Halil. Published in the Ἀμάλθεια, April 27–29, 1884, No. 489." *See also Lightfoot, Apostolic Fathers, Part II. Vol. I. p. 617, note.*

ΔΙΙΛΑΡΑΣΙ
ΩΣΕΒΑΣΤΩ
ΕΥΜΕΝΕΙΚΛΑΥ
ΔΙΩΣΜΕΛΙ
ΤΩΝΟΙΕΡΕΥΣ
ΑΠΟΚΑΤΕ
ΣΤΗΣΕΝ

Διὶ Λαρασί-
ῳ Σεβαστῷ
Εὐμενεῖ Κλαύ-
δι(ο)ς Μελί,
των ὁ ἱερεὺς
ἀποκατέ-
στησεν.

Lightfoot, *loc. cit.* (see also Vol. I. p. 444), points out that the Emperor *Hadrian* is here identified with *Zeus Larasios*, the patron God of Tralleis, and that *Claudius Melito* is perhaps the same person mentioned in *Papers of the American School*, Vol. I. pp. 100, 102, 103, which last corresponds with Le Bas-Waddington, *Voyage Archéologique*, 1652 *c*.

No. 382.

Tralleis. Forwarded to me in February, 1886, *by Mr. Mich. Pappa Konstantinou.* "On a quadrangular cippus of Breccia, found in the house of Mehemet Effendi in Merkeme Mahalesi; published in the 'Αμάλθεια, October 18–30, 1885, No. 860." *See* Bulletin de Correspondance Hellénique, 1886, *p.* 515.

ΩΙΧΑΙΡΕΙΝΑΡΙΣΤΕΑ
ΛΟΕΓΕΓΡΑΦΕΙΤΕΥΠΕ

ΜΑΤΙΓΕΓΡΑΜΜΕΝΟΙΣ
ΧΟΥΠΕΡΙΟΡΙΣΜΟΥΣ
5 ΟΜΕΝΩΝΔΕΚΑΤΗΝΑ
ΔΙΥΜΑΣΠΟΕΙΝΠΑΝΤ
ΠΟΛΛΩΝΙΑΙΣΥΝΤΟΙΣ
ΒΑΣΙΛΙΚΟΝΔΕΚΑΤΗΝΤΩ
ΜΙΣΤΟΚΛΕΙΤΩΙΣΤΡΑΤΗΓ
10 ΡΟΓΕΓΡΑΜΜΕΝΩΝΚΑ

[Βασιλεὺς ὁ δεῖνα τῆι βουλῆι καὶ τῶι δήμ-]
ωι χαίρειν. Ἀριστέα [. . . . . ἀλ-]
[λ]ο ἐγεγρά[φ]ειτε ὑπὲ[ρ . . . . . ἐν τῶι γράμ-]
[μ]ατι γεγραμμένοις [. . . . . Ἀν-]
[τιό]χου περιορισμοὺς . . . .
5 [ο]μένων δεκάτην ἀ . . . .
δι᾽ ὑμᾶς ποεῖν πάντ[α . . . . Ἀ-]
πόλλωνι αἱ σὺν τοῖς . . . .
βασιλικὸν δεκάτην τω[. . . . . . . Θε-]
μιστοκλεῖ τῶι στρατηγ[ῶι . . . . ᾿ . . .]
10 . . γεγραμμένων κα[. . . . . , . . .]

Like No. 4 this inscription is a fragment of a letter of Antiochus
(? see No. 4, lines 3-4), king of Syria, in regard to the people of
Hiera Kome and the sanctuary of Apollo.

No. 383.

*Tralleis. Forwarded to me in October, 1886, by M. Mich.
Pappa Konstantinou.* "Quadrangular block near the house
of Ibrahim Aga." *See* Bulletin de Correspondance Hellénique, 1887, p. 218.[1]

---

[1] Ligatures occur in line 4, MH; line 7, HN; line 12, NH.

```
▨ΒΟΥΛΗΚΑΙΟΔΗ
 ΜΟΣΚΑΙΤΟΙΕΡΟΝ
▨ΥΣΤΗΜΑΤΗΣΓΕ
 ΟΥΣΙΑΣΕΤΙΜΗΣΑΝ
5 ΛΟΥΚΙΛΙΑΝ▨ΓΛΟΥ ▨
 ▨ΙΟΥΘΥΓΑΤΕΡΑΛΑΥ
 ▨ΙΚΗΝΤΗΝΑΡΧΙΕΡΕΙ
 ▨ΝΔΙΑΤΑΣΤΟΥΓΕ
 ΝΟΥΣΑΥΤΗΣΦΙΛΟΤΙ
10      ΜΙΑΣ
 Π̣ΑΙΛΙΟΣΒΑΣΣΟΣΧΡΥ
 ΣΕΡΩΣΣΤΕΦΑΝΗΦΟΡΙ
 ΣΑΣΑΡΧΙΕΡΑΤΕΥΣΑΣ
 ΚΑΙΤΑΣΛΟΙΠΑΣΑΡΧΑΣ
15 ΚΑΙΛΙΤΟΥΡΓΙΑΣΤΕΛΕΣΑΣ
 ΤΗΝΕΑΥΤΟΥΜΗΤΕΡΑ
```

['Η] βουλὴ καὶ ὁ δῆ.
μος καὶ τὸ ἱερὸν
[σ]ύστημα τῆς γε-
[ρο]υσίας ἐτίμησαν

5 Λουκιλίαν Γ(αΐου) Λου[κ]ι.
[λ]ίου θυγατέρα Λαυ-
[δ]ίκην τὴν ἀρχιέρει-
[α]ν διὰ τὰς τοῦ γέ-
ους αὐτῆς φιλοτι-

10   μίας·
Πό(πλιος) Αἴλιος Βάσσος Χρυ-
σέρως στεφανηφορ[ή-]
σας, ἀρχιερατεύσας
καὶ τὰς λοιπὰς ἀρχὰς

15 [κ]αὶ λιτουργίας τελέσας
τὴν ἑαυτοῦ μητέρα.

Concerning the name Λανδίκη, see No. 27.

No. 384.

Tralleis. Forwarded to me in October, 1886, *by M. Mich. Pappa Konstantinou.* "Near the house of Joannes Minaretoghlu."

ΟΙΕΝΤΗΑΞΙΑΔΗΜΟΙΕΤΜΗΞΑΝ
ΔΗΜΗΧΑΙΡΕΜΟΝΟΞΚΑΛΛΙΝΟΗΝ
ΓΕΝΟΜΕΝΗΝΙΕΡΗΑΝΤΗΞΑΡΤΕΜΙ
ΔΟΞΥΠΟΤΟΥΔΗΜΟΥΤΩΝΕΦΕ
ΞΙΩΝ

Οἱ ἐν τῇ 'Ασίᾳ δῆμοι ἐτ[ί]μησαν
Δημῇ Χαιρέμονος Καλλινόην
γενομένην ἱέρ[ει]αν τῆς 'Αρτέμι-
δος ὑπὸ τοῦ δήμου τῶν 'Εφε-
σίων.

No. 385.

Tralleis. Forwarded to me in February, 1886, *by Mr. Mich. Pappa Konstantinou.* "On a quadrangular block, whose height is 0.73 m.; width, 0.84 m.; thickness, 0.28 m. Ten lines are erased at the top." *See* Bulletin de Correspondance Hellénique, 1886, *p.* 326.

K
T
T   Ρ[.ΞΤ[.]Ξ
ΑΥΤΟΝΠΑΡΕΧΟΜ
ΤΟΙΞΔΙΑΦΕΡΟΜΕΝΟΙΞ   ΧΙΕ
ΚΑΘΟΛΟΥΤΕΕΠΙΜΕΛΟΜΕ[ΝΟΙΞ]ΠΡΟΞ Ε
ΚΑΙΛΥΞΙΤΕΛΗΠΕΡΙΕΠΟΙΗΞΕΝΤΗΠΑΤΡΙ[ΔΙ
ΑΝΤΕΧΟΜΕΝΟΞΑΕΙΤΗΞΠΑΡΑΤΩΝΠΟΛΙΤΩΝ
ΜΑΝΤΟΥΑΓΑΘΟΥΞΤΩΝΑΝΔΡΩΝΟΠΩΞΚΑΙΟΙ

ΓΙΝΩΝΤΑΙΠΡΟϹΤΟΤΟΝΔΗΜΟΝΕΥΕΡΓΕΤ▨
ΚΑΙϹΤΕΦΑΝΩϹΑΙΑΥΤΟΝ▨ΧΙ▨
ΠΑΤΡΙΔΑϹΤΗϹΑΙΔΕΑΥΤΟΥΤΗΝΕΙΚ▨
Τ▨ΓΕΝΗ▨ΠΑϹΙΝΤΟΝΕΝΤΩΔΩ▨
ϹΑϹΘΑΙΤΗϹΑΝΑΓΓΕΛΙΑΝΤΗϹΕΙΚΟΝ▨
ΤΟΥΔΗΜΟΥΤΩΒΑϹΙΛΕΙΑΝΑΓΡΑΨΑΙ▨

. . . . [ἵνα φανῇ]
[ἴσον ἑ]αυτὸν παρεχόμ[ενος πᾶσιν τοῖς δικαζομένοις καὶ]
τοῖς διαφερομένοις . . . .
καθ᾽ ὅλου τε ἐπιμελομέ[νοις] προσε . . . .
καὶ λυσιτελῆ περιεποίησεν τῇ πατρί[δι . . . .]
ἀντεχόμενος ἀεὶ τῆς παρὰ τῶν πολιτῶν . . . .
μαν τοὺ(ς) ἀγαθοὺς τῶν ἀνδρῶν ὅπως καὶ οἱ . . . . .
γίνωνται πρὸς τὸ τὸν δῆμον εὐεργετ[εῖν . . . .
καὶ στεφανῶσαι αὐτὸν [εἰκόνι] χ[αλκῇ ἀρετῆς ἕνεκεν τῆς
πρὸς τὴν]
πατρίδα, στῆσαι δὲ αὐτοῦ τὴν εἰ[κόνα ἐπὶ στυλίδος
μαρμαρώης]
. . . . . . πᾶσιν τὸν ἐν τῷ δω[. . . . . ποιή-]
σασθαι τὴ[ν] ἀναγγελίαν τῆς εἰκόν[ος τοὺς θυσιάζοντας
ὑπὲρ?]
τοῦ δήμου τῷ βασιλεῖ, ἀναγράψαι [τὸ ψήφισμα τόδε
εἰς στήλην]
[καὶ στῆσαι ἐν ἐπιφανεστάτῳ τόπῳ?]

No. 386.

Tralleis. Forwarded to me in April, 1886, by Mr. Mich. Pappa Konstantinou. "In the village Acharkieui (one hour distant from Tralleis), in the house Kütchükoghlu Hussein. *See* Bulletin de Correspondance Hellénique, 1886, *p.* 518.

ΤΟΥΠΑΥΤΟΝ
ΜΑΡΚΟΥΑΥΡΗΛΙΟΥ
ΕΤΑΡΑΟΥΚΑΙΓΥΝΑΙ
ΟΥΚΑΙΤΕΚΝΩΝΚΑΙΕΚΤΟ
5   ΡΕΜΜΑΤΩΝΕΛΗΛΥ
ΟΑΥΤΟΝΑΠΟΔΙΑΔΟΧΗΣ
ΔΕΚΝΟΥΧΡΕΓΕΛΛΑΝΙΟΥΟΝΗΣΥ
ΛΟΥΚΑΘΩΣΤΑΕΝΓΡΑΦΑΠΕΡΙΕΧΕΙ
ΖΩΣΙΝ

['Ο βωμὸς καὶ τὸ ὑπ' αὐτὸ]ν
[μνημεῖον] Μάρκου Αὐρηλίου
[γυμνασι]ά[ρχου κ]αὶ γυναι-
[κὸς αὐτοῦ] καὶ τέκνων καὶ ἐκ[γ]ό-
[νων καὶ θ]ρεμμάτων ἐληλυ-
[θότων πρὸς αὐ]τὸν ἀπὸ διαδοχῆς
Δέκ[μ]ου [Φ]ρεγελλανίου 'Ονησύ-
λου, καθὼς τὰ ἔγγραφα περιέχει.
Ζῶσιν.

Var. Lect.

The *Bulletin* marks line 1 as certain; in line 3, *Bulletin* omits
ΡΛΟΥ; in line 4, *Bulletin* marks ΟΥ as certain, and reads ΓΟ at
the end; in line 6, *Bulletin* marks ΑΥ as certain.

No. 387.

Tralleis. Forwarded to me in April, 1886, *by Mr. Mich.
Pappa Konstantinou.* "In the house of Hassan Tchaush,
in Furmali Sokaki (= street); published in the 'Αμάλθεια,
April 27–29, 1884, No. 489." *See* Bulletin de Correspondance Hellénique, 1886, *p.* 455.

▨▨▨AΞIAKAIOΔHMOΣKAIϹ▨▨▨
▨▨▨IONYΣONTEXNITAIETIMH
▨▨▨AΠOΛΛΩNIONΔHMHTPIOY
▨▨▨ONΠPOANAΓEΛENTATHΣΣY
▨▨▨NOΔOYAPXIEPϹ▨▨▨▨

[Τῶν βαφέων ἡ?]
[ἐργ]ασία καὶ ὁ δῆμος καὶ [οἱ]
[περὶ? Δ]ιόνυσον τεχνῖται ἐτίμ[η-]
[σαν] Ἀπολλώνιον Δημητρίου
[τὸ]ν προαναγ[ρα]φέντα? τῆς συ-
[νόδου] ἀρχιερ[έα].

The restoration of lines 1–2 is merely tentative. For ἡ ἐργασία τῶν βαφέων, see *Bulletin de Correspondance Hellénique*, 1886, p. 519. For ἡ συντεχνία τῶν λινύφων, see *Papers of American School at Athens*, I. p. 97. The guilds seem to have been well organized at Tralleis.

No. 388.

Tralleis. Forwarded to me in October, 1886, by M. Mich. Pappa Konstantinou. "On a block of marble in the house of the blacksmith Salih, in the street Tchatal Tcheshme." *The left side is broken away.*[1]

```
            AΞ
      ∠·ϽCΠPOΦACEI
      \IA
      ϽHNAYШPIΣAΘHNA
  5   ΘHNAΓOPOYTOYTE
      ▨TOYNOMIMOYEI▨
      THNΘHKHNΔШΞEI
      \NYΣAΣTШTAMEIШ
      \TEIΔIШNHMШN
```

---

[1] In line 11 NH are in ligature.

10   ⊃ ≤ ¹ \ ≤

ΓΡΑΜΜΑΤΑΟΥΓΕΝΗΣΕΤΑΙ
ΓΡΑΦΟΝΑΠΕΤΕΘΗΕΙΣΤΑ
Ζ Ω Σ Ι Ν

∙ ∙ ∙ ∙ ∙ ∙ ∙ ∙
∙ ∙ ∙ ∙ ∙ ∙ προφάσει
∙ ∙ ∙ ∙ ∙ ∙ ∙ ∙
∙ ∙ ∙ ∙ 'Αθ]ηνᾶ χωρὶς 'Αθηνᾶ[ς]
'Αθ]ηναγόρου τοῦ τε-
∙ ∙ ∙ ∙ τοῦ νομίμου εἰ-
∙ ∙ ∙ ∙ τὴν θήκην δώσει
∙ ∙ ∙ ∙ ∙ ∙ ∙ τῷ ταμείῳ
∙ ∙ ∙ ∙ ∙ ∙ εἰδίων ἡμῶν
∙ ∙ ∙ ∙ ∙ ∙ ∙ ∙
∙ ∙ ∙ γράμματα οὐ γενήσεται ∙
[Τούτου ἀντί]γραφον ἀπετέθη εἰς τὰ
[ἀρχεῖα].
                                    Ζῶσιν.

No. 389.

Tralleis. Forwarded to me in February, 1886, *by M. Mich.
Pappa Konstantinou.* See Bulletin de Correspondance
Hellénique, 1886, *p.* 456.

▨▨▨Ι ΕΝΑΝΔΡΟΝΔΙΤΤΟΝ
▨▨▨ΝΑΣΙΑΡΧΗΣΑΝΤΑΠΑ
▨▨▨ΤΟΥΚΑΙΣΤΡΑΤΗΓΗΣΑΝ
          ΤΑΤΗΣΠΟΛΕΩΣ

∙ ∙ ∙ ∙ [Μ]ένανδρον, δὶ(ς) τὸν
∙ ∙ ∙ ∙ ∙ ∙ ἀσιαρχήσαντα πα-
∙ ∙ ∙ ∙ ∙ ∙ του καὶ στρατηγήσαν-
          τα τῆς πόλεως.

No. 390.

*Tralleis. Forwarded to me in February, 1886, by M. Mich.
Pappa Konstantinou.*

ΘΕΩΝΕΥϽΠΟΛΕϽϹ▨▨▨▨
ΑΥΡΗΛΙΑϹΕΥΦΡΟϹΥΝΕΑΕ▨▨
ΓΡΑΜΜΑΤΕΥϹΑϹ▨ΤΟΥΤΟ▨▨
▨▨▨ΚΑΘΕΙΔΡΥϹΕ▨ΤΟΝϽΕ▨

θεῶν . . . . πόλε[ω]ς
Αὐρηλίας Εὐφροσυν[η . . . .
γραμματεύσας . . . . τουτο
. . . . καθείδρυσε τὸν.

No. 391.

*Tralleis. Forwarded to me in February, 1886, by M. Mich.
Pappa Konstantinou.*

ϽϵΤΕΙΜΑΡΧϹ▨
ϽΠΩΛΗϵ
ΥΝΗΗΓΗϵΙΠΙ▨
ϵΚΑΙΧΡΗϵΤ
ΧΑΙΡΕΙ

. . . . Τείμαρχ[ον . . . .
. . . . πωλης
. . . . γυνὴ Ἡγησίπ[πη] . . . .
[ἄλυπε?] καὶ χρηστ[ὲ]
χαῖρε⟨ι⟩?

No. 392.

*Tralleis.   Forwarded to me in April, 1886, by M. Mich.
Pappa Konstantinou.*

```
         ///////////////A Γ//////////////
         ////A I Π O Λ//////////////
         ///)N Δ A M ////////////
         //FΩTONA I//////////
    5     K A I T H Γ Y N/////////
          Φ I Λ A ///////////
          Δ I A T A Y Π E P//////
          T O Π Λ H Θ O  ////////
          X A I T E K N Ω N////////
    10    Π P O X P E I Δ I K/////
         ///////////.T T O I ///////
```

4 [πρ]ῶτον α
   καὶ τῆς γυν[αικὸς 'Αρετα-?]
   φίλας . . . .
   διὰ τὰς ὑπὲρ . . . .
   τὸ πλῆθος
        τέκνων.

No. 393.

*Tralleis.   Forwarded to me in April, 1886, by M. Mich.
Pappa Konstantinou.* "In the house of Kirli Hussein.
Quadrangular block."

```
          E T H  I A N//////////
          A Γ A Θ I N O//////////
          E  T H  E//////////
          M A N Δ P Ω N//////////
          A P T E M I Δ Ω P O N////
```

ἐτησίαν
ἀγαθ(ὴ)ν [. . . ἀν-]
ἔστησε
ἀνδρῶν
'Αρτεμίδωρον.

No. 394.

Tralleis. Forwarded to me in April, 1886, *by M. Mich. Pappa Konstantinou.* "On a piece of marble built into a wall in the place called *Dede Kuyusu.*" Bulletin de Correspondance Hellénique, 1886, *p.* 327.

```
N I K H Σ A N T A A N Δ P A Σ Δ C Λ I
K O Φ I A Σ I A T Λ░N Y Λ E I T O
Φ I Λ O T E K N O Σ H P Ω Δ O Y Σ N I M I C Σ E Π
```

No. 395.

Tralleis. Forwarded to me in April, 1886, *by M. Mich. Pappa Konstantinou.* "The most of the inscription has been hewn away by a stonemason, the following letters alone being left."

```
Π P O Π
M A Ξ I I
K P A T
Y Π A T
Π P E Σ
Y I O N Λ
M E Σ Σ Λ
Λ A M I T
    A N Θ Y
```

προπ . . . .
Μαξι[μ . . . αὐτο-]
κρατ[ορ . . . . . .
ὑπατ[ο . . . . . .
πρεσ[β . . . . .
υἱὸν . . . . . .
Μεσσ[αλ . . . .
λαμιτ . . . . . .
. . . ἀνθυ[πατ . . .]

No. 396.

Nysa. Forwarded to me in April, 1886, *by M. Mich. Pappa Konstantinou.* "On a quadrangular block, which was brought from *Sultan Hissar* and is now in *Aktchekieui*, one hour distant from *Sultan Hissar;* published in the 'Αμάλθεια, January 26–27, 1884, No. 426." Bulletin de Correspondance Hellénique, 1886, *p.* 454.[1]

```
    ΗΒΟΥΛΗΚΑΙΟΔΗΜΟϹ
    ΕΤΙΜΗϹΑΝΚ·ΚΑΙΚΙΛΙΟΝ
    ΗΡΑΚΛΕΙΔΗΝΝΕΩΤΕΡΟΝ
    ΑΝΔΡΑΑΓΑΘΟΝΗΡΩΑΔΙΑΤΕ
 5  ΗΘΟϹΚΑΙΠΑΙΔΕΙΑΝΚΑΙΤΑϹ
    ΕΚΠΡΟΓΟΝΩΝΕΙϹΤΗΝΠΑ
    ΤΡΙΔΑΦΙΛΟΤΙΜΙΑϹΑΞΙΟΝ
    ΠΑϹΗϹΤΙΜΗϹΑΝΑϹΤΗ
    ΛΑΝΤΟϹΤΟΝΑΝΔΡΙΑΝΤΑ
10  ΚΑΙΚΙΛΙΟΥΕΥΤΥΧΟΥϹΤΟΥ
    ΘΡΕΨΑΝΤΟϹΑΥΤΟΝΑΝΤΙ
    ΤΗϹΙΔΙΑϹΑΥΤΟΥΤΗϹΠΑ
```

---

[1] Ligatures occur in lines: 4, TE; 11, NT.

NHΓΥΡΙΑΡΧΙΚΗϹΤΙΜΗϹΚΑ
ΘΩϹΚΑΙΗΒΟΥΛΗϹΥΝΕ
15   ΧΩΡΗϹΕΝ

'Η βουλὴ καὶ ὁ δῆμος
ἐτίμησαν Κ(όϊντον) Καικίλιον
'Ηρακλείδην νεώτερον
ἄνδρα ἀγαθὸν ἥρωα διά τε
5 ἦθος καὶ παιδείαν καὶ τὰς
ἐκ προγόνων εἰς τὴν πα-
τρίδα φιλοτιμίας, ἄξιον
πάσης τιμῆς, ἀναστή-
σαντος τὸν ἀνδριάντα
10 Καικιλίου Εὐτύχους τοῦ
θρέψαντος αὐτὸν ἀντὶ
τῆς ἰδίας αὐτοῦ τῆς πα-
νηγυριαρχικῆς τιμῆς κα-
θὼς καὶ ἡ βουλὴ συνε-
.15   χώρησεν.

No. 397.

Nysa. Forwarded to me in April, 1886, by *M. Mich. Pappa Konstantinou.* "Built into the wall of a Turkish Djami; published in the 'Aμάλθεια, January 26–27, 1884, No. 426." *See* Bulletin de Correspondance Hellénique, 1886, *p.* 520.[1]

ΑΥΡΗΛΙΑΗΟΝ
ΑΠΕΛΛΑΝ
ΧΕΙΛΙΑΡΧΟΝ
ΛΕΓΙΩΝΟϹΤΡΙ

---

[1] Line 5, HN are in ligature.

5 Τ Η Ξ Κ Υ Ρ Η Ν Α Ι Κ Η Ϲ
   Τ Ο Ν Ε Α Υ Τ Ο Υ
   Π Α Τ Ε Ρ Α
   Κ Α Τ Α Λ Ι Π Ο Ν Τ Α Τ Η
   Ι Ε Ρ Ω Τ Α Τ Η Β Ο Υ Λ Η
10 Ε Ι Ϲ Κ Λ Η Ρ Ο Υ Ϲ Ε Α Υ Τ Ο Υ Κ Α Ι
   Τ Ο Υ Υ Ι Ο Υ Α Π Ε Λ Λ Α ✳ ʃΖ

   Αὐρηλια[νὸν]
   ᾿Απελλᾶν
   χειλίαρχον
   λεγιῶνος τρί-
5 της Κυρηναϊκῆς
   τὸν ἑαυτοῦ
   πατέρα
   καταλιπόντα τῇ
   ἱερωτάτῃ βουλῇ
10 εἰς κλήρους ἑαυτοῦ καὶ
   τοῦ υἱοῦ ᾿Απελλᾶ (δηνάρια) ,ζ.

   Var. Lect.

Line 1, *Bulletin* reads ΑΝΟ ; line 11, end, *Bulletin* reads ΛΛΑ · Ζ.

No. 398.

Nysa. In December, 1886, *M. Mich. Pappa Konstantinou wrote to me concerning an inscription now in the village Aktche, near Nysa, as follows:*

Εἴς τινα Τουρκικὴν οἰκίαν ἀνεκάλυψα ἕν ψήφισμα ΝΥΣΑΕΩΝ ἐκ τριάκοντα καὶ πλέον γραμμῶν ἀναφερόμενον εἰς τὴν ἐποχὴν ΠΥΘΟΔΩΡΟΥ τινος, ἴσως αὐτὸν ὄν ἀναφέρει ὁ Στράβων (12. 555 and 14. 649. See also Cicero, *pro Flacco*, 22, 52 : Ubi erant illi Pythodoro, Archidemi, Epigoni, ceteri homines apud nos noti, inter suos nobiles? Cf. *Papers of the American School at Athens*, I. p. 96). Ἡ ἐπίρροια τῆς ἀτμοσφαίρας ἔχει βλάψει τὴν ἐπιφάνειαν τῆς ἐπιγραφῆς καὶ δι' ἐμὲ ἡ ἀντιγραφὴ καθίσταται δύσκολον ἄνευ σφαλμάτων.

I mention this inscription here in the hope that some traveller may be induced to hunt it up, and secure a copy before it be totally defaced by the weather.

ERRATA.

The following errors have been found in the WOLFE EXPEDITION
TO ASIA MINOR : —

- No. 23, line 1, read (Σ)ύμμαχ(ο)ν
  No. 26, line 2, read Ἀμοῦκιν
  No. 47, line 1, read Παππᾷ
  No. 50, line 5 end, read Α[ὐτ]ο-
  No. 57, line 1, read Ναννᾶς ; and in line 6, read Ἀππᾶς
  No. 59, line 1, read Ναννᾶ
  Page 47, line 9, for 72 read 68.
  No. 91, line 3 init., read ἧς
  No. 97, note, read Σβηνῶμις
  No. 123, line 3, read Νῆνιν
  No. 141, line 1, read Ὁ δεῖνα
  No. 144, line 7, read τ[α]λασίφρονι
  No. 150, line 1, read ἀνέστη-] ; and in line 5, read μν(ήμης)
  No. 174, line 3, read μ[ήποτε?]
  No. 178, note, read *Palaea Isaura.*
  No. 187, line 3, read Ἀρμενι[ακοῦ]
  No. 190, line 6, read ἐπ[α]ρ- ; and in line 10, read ε[ὐε]ργέτην
  No. 269, line 3, read ρ[α-]
  No. 284, line 2, read [ρκιζόμεθα δὲ] Μῆνα καταχθόνιον εἰς
  Page 190, line 6, read Kizil Ören Dagh ; and in line 11, read
  Kötchkieui,
  No. 317, line 1, read Ἀναβουρέων ; and in line 7, read εὐ]εργέτην
  Page 194, line 13, read Εὐβοσία ; and instead of 337 read 339.
  No. 320, line 6, read [τέκ]νοις
  No. 322, line 2, read χαλκε[ὺς]
  No. 332, line 1, read ἱερέ[ων]
  No. 335, line 2, read κρ[άτ]ους

Page 213, *D*, line 7, put a colon at the end of the line.

Page 214, line 21, read οὐρανίας.

Page 214, line 27, put a colon at the end of the line.

Page 214, line 32, read ἐξεῖται, and put a colon at the end of the line.

No. 343, line 4, read Ἀλεξ[ά]νδ-

No. 345, line 2, read Στά-

No. 351, read ἀνεικήτοις

No. 352, after *Yalowadj-Sofular* insert the words (*Antiochia Pisidiae*).

No. 365, line 12, read *Ciminiae*.

No. 366, line 14, read δόντ[ος; in line 16, read δόντος; in line 80, read Τίτος

No. 373, line 37, read Καλου(ή)νιος

No. 375, line 11, read Σουσίλου

No. 376, line 7, read Ὀλυνποκ[ω-]

No. 380, line 11, read ἀν[έ]στη-

Page 272, read Ναζουλεύς

No. 399, line 2, read [αν]οῦ Ἀδριανοῦ

No. 401, line 5, read σύνης

Nos. 404–405, line 1, read ὁ δῆμος

No. 409, uncial text, line 1, read ΘΕΟΙΣ

No. 417, line 10, read Δαρεῖος

No. 429, line 1, read ἐτίμη-

No. 443, line 1, read Αὐρῆλις

No. 444, line 4, read [μνή-]

No. 449, line 5, read [ἐ]τέλεσ[α]ν

No. 468, line 3, read [Μ]ενέ[μα]χος

No. 472, line 13, read πατρικ[ῆ].

Page 342, *C*, line 5, read ϕ

No. 489, line 7, read [γλ]υκυ[τά-]

No. 499, line 6, read εἰδίᾳ

No. 517, after the words *Ulu Borlu* insert the word (*Apollonia*); and in line 11 init., for τῶν read νῶν

No. 548, line 4, read Τυμβριαδέ(α)s

No. 550, line 1, read [θυ-]

No. 570, line 2, read μνή-

No. 609, line 4, read Τατίᾳ

Page 429, instead of 353 read 354; instead of 354 read 355; instead of 355 read 356.

Page 430, instead of 368 read 369; instead of 369 read 370; instead of 372 read 373.

No. 633, read *About one*

No. 640, line 1, read αἰώνιον

Page 448, in the fourth line from the bottom insert the word *way* after the word *some*.

No. 275, lines 7–8, read πεπαῦσθαι

ARCHÆOLOGICAL INSTITUTE OF AMERICA.

AMERICAN SCHOOL OF CLASSICAL STUDIES AT ATHENS.

January, 1888

AMERICAN SCHOOL OF CLASSICAL STUDIES AT ATHENS.

1887–1888.

TRUSTEES.

A corporation was formed in March, 1886, under the statutes of the Commonwealth of Massachusetts, with the name of "The Trustees of the American School of Classical Studies at Athens," to hold the title to the land and building in Athens belonging to the School, and to hold and invest all permanent funds which may be received for its maintenance.

The Board consists of the following gentlemen : —

MANAGING COMMITTEE.

THOMAS D. SEYMOUR, Yale University, New Haven, Conn., *Chairman.*

H. M. BAIRD, University of the City of New York, New York.

I. T. BECKWITH, Trinity College, Hartford, Conn.

FRANCIS BROWN, Union Theological Seminary, 1200 Park Ave., New York.

MISS A. C. CHAPIN, Wellesley College, Wellesley, Mass.

MARTIN L. D'OOGE, University of Michigan, Ann Arbor, Mich.

HENRY DRISLER, Columbia College, 48 West 46th St., New York.

O. M. FERNALD, Williams College, Williamstown, Mass.

A. F. FLEET, University of Missouri, Columbia, Mo.

BASIL L. GILDERSLEEVE, Johns Hopkins University, Baltimore.

WILLIAM W. GOODWIN, Harvard University, Cambridge, Mass., *Chairman of Committee on Publications.*

WILLIAM G. HALE, Cornell University, Ithaca, N. Y.

ALBERT HARKNESS, Brown University, Providence, R. I.

THOMAS W. LUDLOW, Yonkers, N. Y., *Secretary.*

AUGUSTUS C. MERRIAM, Columbia College, New York; *Director of the School* (1877–1888), Athens, Greece.

CHARLES ELIOT Norton (*ex officio*), Harvard University, Cambridge, Mass., *President of the Archæological Institute of America.*

FRANCIS W. PALFREY, 255 Beacon St., Boston.

WILLIAM PEPPER, University of Pennsylvania, 1811 Spruce St., Philadelphia.

FREDERIC J. DE PEYSTER, 7 East 42d St., New York, *Treasurer.*

WILLIAM M. SLOANE, College of New Jersey, Princeton, N. J.

FITZGERALD TISDALE, College of the City of New York, New York.

WILLIAM S. TYLER, Amherst College, Amherst, Mass.

JAMES C. VAN BENSCHOTEN, Wesleyan University, Middletown, Conn.

WILLIAM R. WARE, Columbia College, School of Mines, New York.

JOHN WILLIAMS WHITE, Harvard University, Cambridge, Mass.

EXECUTIVE COMMITTEE.

THOMAS D. SEYMOUR, *Chairman.*	CHARLES ELIOT NORTON.
WILLIAM W. GOODWIN.	FREDERIC J. DE PEYSTER, *Treasurer.*
THOMAS W. LUDLOW, *Secretary.*	WILLIAM R. WARE.

JOHN WILLIAMS WHITE.

ANNUAL DIRECTORS.

1882-1888.

WILLIAM WATSON GOODWIN, Ph.D., LL.D., Eliot Professor of Greek Literature in Harvard University. 1882–83.

LEWIS R. PACKARD, Ph.D., Hillhouse Professor of Greek in Yale University. 1883–84.

JAMES COOKE VAN BENSCHOTEN, LL.D., Seney Professor of the Greek Language and Literature in Wesleyan University. 1884–85.

FREDERIC DE FOREST ALLEN, Ph.D., Professor of Classical Philology in Harvard University. 1885–86.

MARTIN L. D'OOGE, Ph.D., Professor of Greek in the University of Michigan. 1886–87.

AUGUSTUS C. MERRIAM, Ph.D., Professor of Greek in Columbia College. 1887–88.

CO-OPERATING COLLEGES.

1887-1888.

AMHERST COLLEGE.
BROWN UNIVERSITY.
COLLEGE OF THE CITY OF NEW YORK.
COLLEGE OF NEW JERSEY.
COLUMBIA COLLEGE.
CORNELL UNIVERSITY.
DARTMOUTH COLLEGE.
HARVARD UNIVERSITY.
JOHNS HOPKINS UNIVERSITY.

TRINITY COLLEGE.
UNIVERSITY OF THE CITY OF NEW YORK.
UNIVERSITY OF MICHIGAN.
UNIVERSITY OF MISSOURI.
UNIVERSITY OF PENNSYLVANIA.
WESLEYAN UNIVERSITY.
WELLESLEY COLLEGE.
WILLIAMS COLLEGE.
YALE UNIVERSITY.

THE AMERICAN SCHOOL OF CLASSICAL STUDIES
AT ATHENS.

THE American School of Classical Studies at Athens, founded by the Archæological Institute of America, and organized under the auspices of some of the leading American Colleges, was opened October 2, 1882. During the first five years of its existence it occupied a hired house on the 'Οδὸς 'Αμαλίας in Athens, near the ruins of the Olympieum. A large and convenient building has now been erected for the School on a piece of land, granted by the generous liberality of the Government of Greece, on the southeastern slope of Mount Lycabettus; adjoining the ground already occupied by the English School. This permanent home of the School, built by the subscriptions of its friends in the United States, will be ready for occupation early in 1888. During the first months of 1887–88, the School has been accommodated in temporary quarters in the city.

The new building contains the apartments to be occupied by the Director and his family, and a large room which will be used as a library and also as a general reading-room and place of meeting for the whole School. A few rooms in the house are intended for the use of students. These will be assigned by the Director, under such regulations as he may establish, to as many members of the School as they will accommodate. Each student admitted to the privilege of a room in the house will be expected to undertake the performance of some service to the School, to be determined by the Director ; such, for example, as keeping the accounts of the School, taking charge of the delivery of books from the Library and their return, and keeping up the catalogue of the Library.

The Library now contains about 1,500 volumes, exclusive of sets of periodicals. It includes a complete set of the Greek classics, and the most necessary books of reference for philological, archæological, and architectural study in Greece.

The advantages of the School are offered free of expense for tuition to graduates of the Colleges co-operating in its support, and to other American students who are deemed by the Committee of sufficient promise to warrant the extension to them of the privilege of membership. It is hoped that the Archæological Institute may in time be supplied with the means of establishing scholarships, which will aid some members in defraying their expenses at the School. In the mean time, students must rely upon their own resources, or upon scholarships which may be granted them by the Colleges to which they belong. The amount needed for the expenses of an eight months' residence in Athens differs little from that required in other European capitals, and depends chiefly on the economy of the individual.

A peculiar feature of the temporary organization of the School during its first six years, which has distinguished it from the older German and French schools at Athens, has been the yearly change of Director. This arrangement, by which a new Director has been sent out each year by one of the co-operating Colleges, was never looked upon as permanent ; and it has now been decided to begin the next year (1888–89) with a new organization. A Director will henceforth be chosen for a term of five years, while an Annual Director will also be sent out each year by one of the Colleges to assist in the conduct of the School. (See Regulation V.) Dr. CHARLES WALDSTEIN, of New York, now Director of the Fitzwilliam Museum of Art at the University of Cambridge, England, has been chosen Director of the School for five years beginning in October, 1888 ; and he has accepted the appointment on the condition that a sufficient permanent fund be raised before that time to support the School under its new organization. It is therefore earnestly hoped and confidently expected that the School will henceforth be under the control of a permanent Director, who by continuous residence at Athens will accumulate that body of local and special knowledge without which the highest purpose of such a school cannot be fulfilled. In the mean time the School has been able, even under its temporary organization, to meet a most pressing want, and to be of some service to classical scholarship in America. It has sought at first, and it must continue to seek for the present, rather to arouse a lively interest in classical archæology in American Colleges than to accomplish distinguished achievements. The lack of this interest has heretofore been conspicuous ;

but without it the School at Athens, however well endowed, can never accomplish the best results. A decided improvement in this respect is already apparent; and it is beyond question that the presence in many American Colleges of professors who have been resident a year at Athens under favorable circumstances, as annual directors or as students of the School, has done much, and will do still more, to stimulate intelligent interest in classic antiquity.

REGULATIONS OF THE AMERICAN SCHOOL OF CLASSICAL STUDIES AT ATHENS.

I. The object of the American School of Classical Studies is to furnish an opportunity to study classical Literature, Art, and Antiquities in Athens, under suitable guidance, to graduates of American Colleges and to other qualified students; to prosecute and to aid original research in these subjects; and to co-operate with the Archæological Institute of America, so far as it may be able, in conducting the exploration and excavation of classic sites.

II. The School is in charge of a Managing Committee. This Committee, which was originally appointed by the Archæological Institute, disburses the annual income of the School, and has power to add to its membership and to make such regulations for the government of the School as it may deem proper. The President of the Archæological Institute and the Director and the Annual Director of the School are *ex officio* members of the Managing Committee.

III. The Managing Committee meets semi-annually, in New York on the third Friday in November, and in Boston on the third Friday in May. Special meetings may be called at any time by the Chairman.

IV. The Chairman of the Committee is the official representative of the interests of the School in America. He presents a report annually to the Archæological Institute concerning the affairs of the School.

V. 1. The School is under the superintendence of a Director. The Director is chosen and his salary is fixed by the Committee.

The term for which he is chosen is five years. The Committee provide him with a house in Athens, containing apartments for himself and his family, and suitable rooms for the meetings of the members of the School, its collections, and its library.

2. Each year the Committee appoints from the instructors of the Colleges uniting in the support of the School an Annual Director, who resides in Athens during the ensuing year and co-operates in the conduct of the School. In case of the illness or absence of the Director, the Annual Director acts as Director for the time being.

VI. The Director superintends personally the work of each member of the School, advising him in what direction to turn his studies, and assisting him in their prosecution. He conducts no regular courses of instruction, but holds meetings of the members of the School at stated times for consultation and discussion. He makes a full report annually to the Managing Committee of the work accomplished by the School.

VII. The school year extends from the first of October to the 1st of June. Members are required to prosecute their studies during the whole of this time in Greek lands under the supervision of the Director. The studies of the remaining four months necessary to complete a full year (the shortest time for which a certificate is given) may be carried on in Greece or elsewhere, as the student prefers.

VIII. Bachelors of Arts of co-operating Colleges, and all Bachelors of Arts who have studied at one of these Colleges as candidates for a higher degree, are admitted to membership in the School on presenting to the Committee a certificate from the instructors in Classics of the College at which they have last studied, stating that they are competent to pursue an independent course of study at Athens under the advice of the Director. All other persons desiring to become members of the School must make application to the Committee. Members of the School are subject to no charge for instruction. The Committee reserves the right to modify the conditions of membership.

IX. Each member of the School must pursue some definite subject of study or research in classical Literature, Art, or Antiquities, and must present a thesis or report, embodying the results of some important part of his year's work. These theses, if approved by the Director, are sent to the Managing Committee, by which each thesis is referred to a sub-committee of three members, of whom two are appointed by the Chairman, and the third is always the Director under whose supervision the thesis was prepared. If recommended for publication by this sub-committee, the thesis or report may be issued in the Papers of the School.

X. When any member of the School has completed one or more full years of study, the results of which have been approved by the Director, he receives a certificate stating the work accomplished by him, signed by the Director of the School, the President of the Archæological Institute, and the Chairman and the Secretary of the Managing Committee.

XI. American students resident or travelling in Greece who are not regular members of the School may, at the discretion of the Director, be enrolled as special students and enjoy the privileges of the School.

PUBLICATIONS OF THE AMERICAN SCHOOL OF CLASSICAL
STUDIES AT ATHENS. 1882–1888.

The Annual Reports of the Committee may be had gratis on application to the Secretary of the Managing Committee. The other publications are for sale by Messrs. Damrell & Upham, 283 Washington Street, Boston.

First, Second, and Third Annual Reports of the Managing Committee, 1881–84. pp. 30.

Fourth Annual Report of the Committee, 1884–85. pp. 30.

Fifth and Sixth Annual Reports of the Committee, 1885–87. pp. 56.

Bulletin I. Report of William W. Goodwin, Director of the School in 1882–83. pp. 33. Price 25 cents.

Bulletin II. Memoir of Lewis R. Packard, Director of the School in 1883–84, with Resolutions of the Committee and the Report for 1883–84. pp. 34. Price 25 cents.

Preliminary Report of an Archæological Journey made in Asia Minor during the Summer of 1884. By J. R. S. Sterrett. pp. 45. Price 25 cents.

PAPERS OF THE SCHOOL.

Volume I. 1882–83. Published in 1885. 8vo. pp. viii. and 262. Illustrated. Price $2.00.

CONTENTS: —

1. Inscriptions of Assos, edited by J. R. S. Sterrett.
2. Inscriptions of Tralleis, edited by J. R. S. Sterrett.
3. The Theatre of Dionysus, by James R. Wheeler.
4. The Olympieion at Athens, by Louis Bevier.
5. The Erechtheion at Athens, by Harold N. Fowler.
6. The Battle of Salamis, by William W. Goodwin.

Volume II. 1883–84. Published in 1888. An Epigraphical Journey in Asia Minor in the summer of 1884, with 397 Inscriptions, mostly hitherto unpublished. By J. R. Sitlington Sterrett, Ph.D. With two Maps, made for this volume by Professor H. Kiepert of Berlin, from the observations and measurements of Dr. Sterrett. 8vo. pp. vii. and 341. Price $2.25.

Volume III. 1884–85. Published in 1888. The Wolfe Expedition to Asia Minor in 1885, with 651 Inscriptions, mostly hitherto unpublished. By J. R. Sitlington Sterrett, Ph.D. With two Maps, made for this volume by Professor H. Kiepert, from the observations and measurements of Dr. Sterrett. 8vo. pp. vii. and 448. Price $2.50.

Volume IV. 1885–86. Published in 1888. 8vo. pp. 277. Illustrated. Price $2.00.

CONTENTS: —

1. The Theatre of Thoricus, Preliminary Report by Walter Miller.
2. The Theatre of Thoricus, Supplementary Report by William L. Cushing.
3. On Greek Versification in Inscriptions, by Frederic D. Allen.
4. The Athenian Pnyx, by John M. Crow; with a Survey of the Pnyx and Notes by Joseph Thacher Clarke.
5. Notes on Attic Vocalism, by J. McKeen Lewis.